CLIMATE CHANGE AND ITS CAUSES, EFFECTS AND PREDICTION

LAND DEGRADATION

THE MAIN CHALLENGE

CLIMATE CHANGE AND ITS CAUSES, EFFECTS AND PREDICTION

Additional books and e-books in this series can be found on Nova's website under the Series tab.

CLIMATE CHANGE AND ITS CAUSES, EFFECTS AND PREDICTION

LAND DEGRADATION

THE MAIN CHALLENGE

ILARIA ZAMBON
LUCA SALVATI
AND
CARLOTTA FERRARA
EDITORS

NOTICE TO THE READER

The Publisher has taken reasonable care in the preparation of this book, but makes no expressed or implied warranty of any kind and assumes no responsibility for any errors or omissions. No liability is assumed for incidental or consequential damages in connection with or arising out of information contained in this book. The Publisher shall not be liable for any special, consequential, or exemplary damages resulting, in whole or in part, from the readers' use of, or reliance upon, this material. Any parts of this book based on government reports are so indicated and copyright is claimed for those parts to the extent applicable to compilations of such works.

Independent verification should be sought for any data, advice or recommendations contained in this book. In addition, no responsibility is assumed by the Publisher for any injury and/or damage to persons or property arising from any methods, products, instructions, ideas or otherwise contained in this publication.

This publication is designed to provide accurate and authoritative information with regard to the subject matter covered herein. It is sold with the clear understanding that the Publisher is not engaged in rendering legal or any other professional services. If legal or any other expert assistance is required, the services of a competent person should be sought. FROM A DECLARATION OF PARTICIPANTS JOINTLY ADOPTED BY A COMMITTEE OF THE AMERICAN BAR ASSOCIATION AND A COMMITTEE OF PUBLISHERS.

Additional color graphics may be available in the e-book version of this book.

Library of Congress Cataloging-in-Publication Data

ISBN: 978-1-53615-575-4

Published by Nova Science Publishers, Inc. † New York

CONTENTS

PREFACE

Desertification is one of the most important issues facing our societies because of its serious consequences for human health, landscape and the environment. Nonetheless, the issue has been in the eyes of media, decision makers and public opinion and it should be noted that this interest tends to be cyclical, corresponding to peaks that reflect the outbreak of emergency situations related to prolonged episodes of drought and water scarcity, in turn associated with climate changes. This volatile interest has focused on the relationship between desertification and climate change (and more generally on the biophysical factors underlying desertification), neglecting the important role played by social, economic, cultural, political and institutional factors. This role — brought to the fore by the most recent socioeconomic dynamics at various spatial scales — requires dedicated approaches from the scientific point of view and a less sensationalistic dissemination of research evidence. This book proposes a trans-disciplinary vision on issues of desertification and land degradation, focusing on long-term socio-ecological dynamics as an interpretative key to local systems' complexity.

In: Land Degradation: The Main Challenge ISBN: 978-1-53615-575-4
Editors: Ilaria Zambon et al.

Chapter 1

INTRODUCTION: (ZERO NET) LAND DEGRADATION - THE MAIN CHALLENGE

***Andrea Colantoni*[1,*], *Ilaria Zambon*[1,†], *Luca Salvati*[2,‡] *and Anastasios Mavrakis*[3,§]**
[1]Tuscia University, Viterbo, Italy
[2]Council for Agricultural Research and Economics (CREA), Arezzo, Italy
[3]Panteion University, Athens, Greece

Desertification is one of the most important issues facing our societies because of its serious consequences for human health, landscape and the environment. Nonetheless the issue has often been the focus of media attention, decision makers and public opinion, it should be noted that this interest tends to be cyclical, corresponding in peaks that reflect the outbreak of emergency situations related to prolonged episodes of drought and water scarcity (Chopra and Gulati, 1997; Salvati et al., 2008a; Colantoni et al., 2015a), in turn associated with climate changes (Barbier,

* Corresponding Author's E-mail: colantoni@unitus.it
† Author's E-mail: ilaria.zambon@unitus.it
‡ Author's E-mail: luca.salvati@crea.gov.it
§ Author's E-mail: mavrakisan@yahoo.gr

2000; Kok et al., 2004; Harte, 2007). This volatile interest has focused on the relationship between desertification and climate change (and more generally on biophysical factors underlying desertification), neglecting the important role played by social, economic, cultural, political and institutional factors (Bajocco et al., 2011). This role - brought to the fore by the most recent socioeconomic dynamics at various spatial scales - requires dedicated approaches from the scientific point of view and a less sensationalistic dissemination of research evidence. By contrast, land degradation would require continuous attention through permanent assessment of ecosystem's quality and stability (Kazemzadeh-Zow et al., 2017; Kairis et al., 2013; Kosmas et al., 2016).

Since land degradation is both a social and economic problem, it is structurally linked to our system of production and consumption (Salvati, 2013, 2014; Rontos et al., 2016). This is based on a concept of progress, which has long been predominant, being centered on the quantitative expansion of production, according to the principle that economic growth would automatically lead to an increase in social well-being. This calls for reflection on our own development model and the need to rethink it in the direction of building "sustainable human well-being" (Arrow et al., 1995). The latter assumes that the improvement of people's living conditions depends not only on increasing availability of goods but also on fair distribution of wealth, reducing the environmental impact of economic growth (Salvati and Carlucci, 2014). At the same time, ensuring stability of welfare over time is also required, based on a concept of intergenerational equity that requires not to compromise the ability of future generations to use the resources that are now available to current generations (Salvati and Zitti, 2008).

Changes toward a different economic model based on sustainable development implies preservation of capital stocks (natural, physical, human and social) from which societies draw resources to build their well-being. In the transition towards a low-carbon economy, in addition to technical and political choices, the role played by local actors who participate in the definition of the decision-making sphere through transformation of their production models, consumption habits and lifestyles, is particularly important. They are called upon to make a cultural leap in overcoming unsustainable models of natural resource consumption and in reducing the environmental impact of human activities (Biasi et al., 2017), so that the ethics of sustainable development becomes culture, political action and economic practice. Cultural change toward a widespread environmental awareness requires the active involvement of scientific and political forces, through democratic participation in land-use decisions.

Sustainability of production processes appears to be an important topic of multidisciplinary research that invests skills in the fields of geography, economics, sociology and environmental science. Environmental pressure deriving from production processes in agriculture, for instance, has led to soil degradation and triggers a progressive loss in biological and economic land potential, impoverishing natural ecosystems and reducing its biodiversity. Land degradation is usually associated with complex environmental conditions that lead, to a generalized loss of environmental quality. These processes, which involve physical and biological aspects of the ecosystem have repercussions on local societies and economic activities being in turn directly (or indirectly) influenced by them. This will produce an environmental spiral that appears increasingly difficult to understand and counteract (Geist and Lambin, 2004).

In advanced economies, drastic changes in social structures reflecting the transition from modernity to a ‘post-modern’ context have occurred over the last few years (Wilson and Juntti, 2005). These processes imply significant changes in the availability of natural resources and landscape configuration with impacts on ecosystems’ stability (Harte, 2007). For instance, changes in the structure of Mediterranean economies since World War II have caused pressures on fragile ecosystems, especially in the most sensitive and economically-disadvantaged areas (Garcia Latorre et al., 2001; Tanrivermis, 2003; Briassoulis, 2005; Salvati and Zitti, 2005, 2012; Atis, 2006; Salvati et al., 2012a). These changes include (i) massive urbanization since the early 1950s, (ii) intensification of agriculture since the 1960s, (iii) economic boom and population growth, particularly during the 1960s and 1970s and (iv) concentration of seasonal tourism with a marked character since the 1980s (Blaikie and Brookfield, 1987; Colantoni et al., 2015a; Cuadrado-Ciuraneta et al., 2017). These changes have attracted increasing attention in both environmental disciplines and social sciences (Puigdefabregas and Mendizabal, 1998). In fact, both disciplines are interested in highlighting the possible consequences of ecosystem degradation on populations and economic systems, as well as the response of societies to such changes (Wilson and Juntti, 2005). Uncertainty and risk are therefore seen as key concepts of this interpretative path that concerns economic dynamics, social changes, cultural development, and political action (Zuindeau, 2007; Salvati et al., 2008b).

Focusing on the primary sector, the environmental problems common to a large part of Mediterranean agriculture are linked to the progressive concentration and specialization of production systems, on the one hand, and to the marginalization of areas less suitable for agriculture, on the other hand (Salvati and Zitti, 2005, 2012; Salvati et al., 2012b; Bajocco et al., 2013; De Rosa and Salvati, 2016). Rural landscapes mix intense cropland with abandoned fields (Salvati and Carlucci, 2011; Colantoni et al., 2015a). The possibility of rebalancing an increasingly unsustainable system is linked to the adoption of eco-compatible production systems in lowlands and to high-quality agricultural productions in marginal areas (Trisorio-Liuzzi and Hamdy, 2002; Salvati et al., 2013c).

Recent studies on agricultural systems have assessed the relevance of different economic, social and environmental dimensions (Trisorio, 2005). The first dimension identifies resource efficiency, the vitality of the agricultural sector and the contribution of the primary sector to conservation of rural areas (Salvati and Carlucci, 2011; Zitti et al., 2015). The social dimension refers to human capital and its characteristics on a local scale (Salvati and Carlucci, 2014; Recanatesi et al., 2016; Rontos et al., 2016; Biasi et al., 2017). The environmental dimension highlights the mechanisms of management and conservation of natural resources, in terms of landscape, water resources and soil (e.g., Salvati et al., 2008a). This framework demonstrates how an integrated and systemic evaluation may provide a correct interpretation of on-going processes, with implications for mitigation and adaption policies (Abraham et al., 2006; Munafò et al., 2013; Smiraglia et al., 2016).

Loss in land productivity triggers soil degradation involving physical, biological and economic aspects of the ecosystem (Brandt, 2005). The notion of soil degradation is linked to desertification, a specific kind of land degradation leading to the irreversible loss of capability for sustainable agriculture and forestry. Different definitions of desertification were proposed according to geographical region, landscape features, spatial scale and territorial characteristics (Lambin, 1993). Climate change, demographic pressure and land-use changes are considered exogenous factors affecting land degradation (Kelly et al., 2015; Karamesouti et al., 2015). The latter are the consequence of an unsustainable exploitation of natural resources (Kosmas et al., 2016), which determines soil depletion, triggering processes that are sometimes irreversible and encouraging the abandonment of farmlands that are no longer productive (D'Angelo et al., 2000). The same areas became economically-disadvantaged and socially-marginal (Kairis et al., 2013, 2014; Karamesouti et al., 2015; Delfanti et al., 2016).

Alongside environmental degradation, the issue of sustainable development is increasingly important and debated worldwide, at the most disparate levels of analysis (Lawn, 2003). The debate on the regional organization of urban and rural regions is particularly intense in the European context, being characterized by a reduced accessibility of inland, hilly and mountainous areas. Since economic, social and the environmental systems are becoming more integrated and interconnected, sustainable development is also becoming increasingly topical at local level. The debate on regional organization is linked to these concepts, with specific relevance in contexts characterized by urban sprawl, which increases land fragmentation (Salvati, 2014; Colantoni et al., 2015a, 2016). If the network of medium-sized cities was demonstrated to produce interesting shares of national wealth, nowadays two directions are identified as promoters of local development: (i) economic functions stimulating innovation and linking new competitive districts to globalization and (ii) a territorial dimension of life quality, linked to areas with a high cultural and environmental image, which also springs from the civic sense of belonging. In this second dimension, the productive territory emerges in wellness services, high-quality and traditional goods, rural hospitality opposed to the paradigms of concentration, scale economies, accessibility and man-made artificial landscapes.

The territorial component of the current socioeconomic scenario is thus key to local development under the assumption that economic investments move from subsidies to engine of a spirit of trust in local resources and entrepreneurship. Local contexts are made even more complex with the drastic reduction in public investment due to budgetary constraints (Cuadrado-Ciuraneta et al., 2017). Heterogeneity in land degradation processes, one of the most typical environmental syndromes (Basso et al., 2000), complicates monitoring and limits the development of effective enforcement actions (Delfanti et al., 2016). However, interest in these issues, particularly around the Mediterranean basin, has been renewed over the last 20 years. Nearly 300 million hectares are affected by soil degradation in Europe (Munafò et al., 2013). These data, derived from national and supranational surveys have contributed to a greater awareness of severity and extent of desertification risk, shrinking the erroneous belief that only the poorest and marginal areas in the world can be affected by land degradation (Salvati et al., 2013a). Even in a developed socioeconomic context, it is possible to recognize desertification problems, sometimes due to climate change, sometimes due to the socioeconomic system, in most cases depending on a complex set of intertwined causes (Zambon et al., 2017, 2018).

At present, the European Mediterranean countries (Portugal, Spain, Italy, Greece, Albania, Bosnia and Herzegovina, Croatia, Cyprus, France, Malta, Slovenia, Spain and Turkey) are particularly exposed to desertification. About 37 million hectares are classified as prone to land degradation (Salvati et al., 2009, 2013b). Critical conditions can be recognized locally in some regions of the above-mentioned countries, where the percentage of the threatened land is even higher, for example in southern Portugal, in regions of Spain and Greece, in Sardinia and Sicily (Di Feliciantonio and Salvati, 2015). Desertification in northern

Mediterranean basin is attributable to a set of causes including biophysical and anthropogenic drivers that generate untenable environmental pressures (Geeson et al., 2002). Early studies have considered land degradation as a complex and composite concept, which describes how one (or more) components of natural capital have deteriorated from a quantitative (or qualitative) point of view (Montanarella, 2007). In recent years, increasing interest in a more systematic investigation of socioeconomic factors that interact with natural capital arises, using both theoretical and empirical approaches (Wilson and Juntti, 2005; Salvati et al., 2008a). However, the relationship between sustainable development and land degradation is still underexplored in the Mediterranean region (Vogt et al., 2011; Kosmas et al., 2016; Pili et al., 2017).

Land degradation in the Mediterranean region reflects gap between growing economies and disadvantaged rural systems, still characterized by internal imbalances. In a territorial context where cultural, social and institutional factors are particularly heterogenous, with efficiency gaps and unequal distribution of natural capital, land degradation results in drastic loss of income, migration and social conflicts (Salvati and Zitti, 2005, 2012; Salvati et al., 2012a). Exposure to land degradation varies considerably over space, widening spatial divides in resource availability and development potential. Vulnerability to land degradation implies the increase of marginal areas with socioeconomic consequences, including rural poverty (Salvati et al., 2011, 2013a; Duvernoy et al., 2018; Zambon et al., 2018). For instance, degraded landscapes often assume the characteristic physiognomy of bad lands, coastlines exposed to soil erosion, abandoned fields in peri-urban areas, marginal landscapes awaiting 're-naturalization' (Conacher and Sala, 1998). The economic activities in these areas, originally based on primary and secondary sectors, often evidence a development path that, as in the case of mass tourism (Cuadrado-Ciuraneta et al., 2017), can upset the delicate environmental balance of rural land (Conacher, 2000).

The environmental consequences of rapid economic development are frequently associated with climate change. A generalized increase in air temperatures has led to a significant increase in soil aridity, which has often been accompanied by changes in rainfall regimes (Dore, 2005). According to IPCC (Intergovernmental Panel on Climate Change) reports, the average surface temperature of the Earth increased by 0.7°C, over one century. However, heating had taken place in a spatially heterogenous manner: the years after 1990 have proved to be the hottest since 1850. According to IPCC estimates, average temperature increases of 0.2 - 0.4°C are expected over the coming 20 years, while in the long-term average temperatures could even increase by another 2 - 4°C (Salvati et al., 2009).

Changes in thermometric regimes are connected to alterations in pluviometric regimes (decrease of annual rainfalls and extreme events), while climate anomalies can be taken as local signals of a more general climate change (Savo et al., 2012). Projections from current climate trends outline future scenarios characterized by higher temperatures and progressively less abundant rainfalls, with implications for soil degradation and water supply (Salvati et al., 2008a). Climate change therefore involves not only purely environmental aspects and the consequences can be quantified in economic terms. Considering climate as a complex system, one can argue how the terms change, variability, anomaly and equilibrium do not concern specific places or variables, implying indeed alterations of more general cause-effect relationships.

To cope with this global emergency, a knowledge-based approach is essential. Informed approaches enable decision-makers to develop the most effective environmental policies and to stimulate the public debate. Knowledge becomes a fundamental prerequisite both for managing land resources and for mitigating soil degradation. In this context, tools assessing the relationship between (changing) economic spaces and complex social systems are especially required. An in-depth analysis identifying some "stylized facts" of the relationship between socioeconomic conditions and land degradation in the Mediterranean basin is proposed in this book. The relationship between economy, society and the environment are interpreted here as a model of participating to a process of 'territorial transition' with individual dimensions contribute (more or less) significantly to sustainable development paths. Path-dependence and the environmental implications of the economic structure lead to different considerations, grounded in the empirical relation between land-use, soil degradation and socioeconomic background (Salvati, 2014; Karamesouti et al., 2015; Salvati et al., 2016).

Although difficult to quantify, the costs resulting from land degradation are considerable even in affected countries and require an accurate assessment over time. Indeed, the adoption of measures to curb soil erosion and degradation is hampered by (i) insufficient and incomplete data on the extent of land degradation and (ii) restricted perception of land operators and their reaction to soil degradation (Kairis et al., 2013, 2015; Ceccarelli et al., 2014; Kosmas et al., 2016). For instance, the effectiveness of economic incentives in influencing farmers' choice to invest in soil protection depends on the immediacy of the benefits. Changes in cultivation practices and landscape protection can only be applied (and give satisfactory results in the mitigation of soil degradation) if operators perceive these processes as a direct threat to land profitability. These measures are also appropriate in areas where agricultural production makes a significant contribution to regional income. National and regional policies and the awareness of local actors, together with a more complete knowledge of the socioeconomic context can contrast current environmental trend through targeted action plans.

Countries in the Mediterranean basin are part of the UNCCD under Annex IV, and include Albania, Bosnia and Herzegovina, Croatia, Cyprus, France, Greece, Italy, Malta, Portugal, San Marino, Slovenia, Spain and Turkey (Corona et al., 2009). The UNCCD produced guidelines to identify and process desertification indicators useful for monitoring land vulnerability in Mediterranean environments (Salvati et al., 2009; 2013c; Kairis et al., 2013). These indicators are based on available data sources, including satellite imagery, topographic data, climatic and geological map, as well as land-use patterns reflecting the impact of socioeconomic aspects (Karamesouti et al., 2015). Annex IV countries will fulfill the objectives of the Convention in terms of both spatial planning and practical intervention. For instance, the National Committee for the Fight Against Desertification in Italy aims to coordinate the implementation of the Convention with the following objectives:

- to identify strategies and priorities in the framework of sustainable development plans and policies to combat desertification and mitigate the effects of drought;
- to prepare and implement a National Action Plan to combat desertification;
- to specify indicators assessing desertification;
- to make an inventory of local technologies, knowledge and traditional practices that contribute to resource saving and the fight against desertification;
- to promote training and research activities;
- to coordinate practical activities with other Mediterranean countries.

Guidelines and the resulting plans are inspired to identification of strategies and objectives that stimulate and re-orient desertification policies. The National Action Plans provide the base for land management with guidelines implementing policies to combat desertification in a context of sustainable development. A permanent monitoring of land ecosystems was established, which promotes the contribution of grassroots communities by stimulating active participation of relevant local authorities. In turn, the joint activity of the Mediterranean countries within the European Commission aims to redirect the interests of the Commission to the problems of land degradation and desertification. Agro-forestry and hydro-geological resource utilization programs are tailored to containment measures and fight against desertification (Cimini et al., 2013; Corona et al., 2014; Marchetti et al., 2015; Biasi et al., 2015; Colantoni et al., 2015b, 2016). From this point of view, complex questions deserve articulated responses from the point of view of measurement and support of spatial policies and local practices. Geographical analysis can interpret the evolution of these phenomena and latent feedback relationships. The restricted availability of historical time series of environmental statistics at municipal scale does not allow the formulation of a broader dataset. Nevertheless, the results obtained appear to be sufficiently detailed and allow a comprehensive assessment of the relationship between land degradation and land organization, supported by empirical evidence arising from the comparison of different socioeconomic contexts.

The empirical results of recent studies highlight critic conditions to land degradation with a spatial pattern changing radically in the most recent period and involving previously unaffected areas (Figure 1). Growing vulnerability to land degradation was progressively decoupled from geographical gradients (Colantoni et al., 2015a; Carlucci et al., 2017; Duvernoy et al., 2018). Its impact on landscape quality reflects a rapidly changing linkage between 'local' and 'global' dimensions of socioeconomic processes.

Elements facing this new "economic-environmental" challenge require integration of different institutional levels and cooperation among actors involved in efficient land management (Oxley and Lemon, 2003). This opens a different phase of stronger relationships with the territory, which becomes the new starting point for tackling sustainable development and desertification risk (Lemon et al., 1994; Salvati and Zitti, 2009; Pili et al., 2017). Earlier research has been designed with the aim of developing indicators and evaluation models to be applied at various spatial scales, based on previous experience from Mediterranean Europe (Ceccarelli et al., 2006; Bajocco et al., 2013; Kairis et al., 2013). Information obtained in this way are intended to meet scientific and management needs through a quantitative assessment of land degradation, its evolution over time, the relationship between land degradation and the economic context.

A set of thematic indicators was proposed to illustrate the main factors of vulnerability predisposing land to degradation. These indicators may exploit the potential of official statistics (General Censuses of Population, Buildings, Agriculture, Industry and Services), contributing to integration with other data sources (e.g., maps from the Corine Land Cover project). These indications are considered a valid reference for different levels of investigation (e.g., national, regional and local scale). Table 1 shows an analytical work plan highlighting critical points, the solutions adopted, and the objectives achieved.

Source: Own elaboration.

Figure 1. Climate aridity and soil quality as contextual indicators of land vulnerability (top left: Filis, Attiki, Greece; right: Diamante, Calabria, Italy; bottom left: Mitikas, Aitolokarnania, Greece; right: Craco, Basilicata, Italy).

According to the research plan, the present work is organized in five chapters. In the first chapter land degradation is framed, through a critical approach to the current conceptual schemes. A detailed description of the underlying causes of land degradation is also proposed, classifying them by simple interpretative criteria, considering the consequences on local economies and societies (Salvati and Carlucci, 2014). The second chapter is devoted to an introduce an interpretative scheme underlying the multifaceted relationship between territorial contexts, sustainable development and land degradation. A review on definitions of sustainable development and the ongoing debate on measuring sustainability is also proposed. This chapter deals with economic development and territorial gaps, referring to EU policy guidelines considering social cohesion and sustainability. In this context, the definition of sustainability is broadened to include economic, social and environmental aspects. Therefore, environmental indicators are analyzed from the point of view of integrated governance and the concept of environmental degradation is exemplified through land degradation processes (Salvati et al., 2012a; Kairis et al., 2013; Delfanti et al., 2016). The third chapter illustrates methodologies and techniques adopted to evaluate land degradation and to analyze the interrelations with the socioeconomic context at local level. Based on specific case studies, the fourth chapter discusses the regional evolution of vulnerability to land degradation at different spatial scales (Smiraglia et al., 2016; Carlucci et al., 2017; Tomao et al., 2017). The discussion also takes stock of responses promoting effective mitigation of land degradation. Policy guidelines, operating tools and intervention practices are proposed according to the local context and, finally, measures mitigating land degradation are investigated using appropriate indicators. Finally, the work concludes with a comment on adaptable policies with reference to Community instruments and their application to environmental issues in ecologically-disadvantaged regions.

Table 1. Research plan and basic operational points

1. State of the art and international research on land degradation and sustainable development; 2. Definition of the problem, terminology adopted, concepts in use, also in relation to their operationalization; 3. Generalization of a logical-conceptual model (DPSIR) for the representation and interpretation of the relationship between land degradation and sustainable development; 4. Identification of the research dimensions in the field of land degradation, understood as critical factors (climate, soil, vegetation, anthropogenic pressure); 5. Choice of environmental variables for analysis and their transformation into vulnerability indicators according to the ESA scheme; 6. Diachronic maps of land vulnerability incorporating long-term scenarios; 7. Selection of socioeconomic variables and their adaptation into quantitative indicators combined with environmental indicators; 8. Mapping of elementary indicators; 9. Summary results commented according to policy indications provided by Annex IV of the World Convention to Combat Drought and Desertification and the National Action Plan to Combat Desertification.

Source: Own elaboration.

References

Abraham, E., Montaña, E. & Torres, L. (2006). Desertification and indicators: possibilities of integrated measurement of complex phenomena. *Scripta Nova. Revista electrónica de geografía y ciencias sociales*, *10*, 214 <http://www.ub.es/geocrit/sn/sn-214.htm> [ISSN: 1138-9788].

Arrow, K., Bolin, B., Costanza, R., Dasgupta, P., Folke, C., Holling, C. S., Jansson, B. O., Levin, S., Mäler, K. G., Perrings, C. & Pimentel, D. (1995). Economic Growth, Carrying Capacity, and the Environment. *Science*, *268*, 520-521.

Atis, E. (2006). Economic impacts on cotton production due to land degradation in the Gediz Delta, Turkey. *Land Use Policy*, *23*, 181-186.

Bajocco, S., De Angelis, A. & Salvati, L. (2012). A satellite-based green index as a proxy for vegetation cover quality in a Mediterranean region. *Ecological Indicators*, *23*, 578-587.

Barbier, E. B. (2000). The economic linkages between rural poverty and land degradation: some evidence from Africa. *Agriculture, Ecosystems and Environment*, *82*, 355-370.

Basso, F., Bove, E., Dumontet, S., Ferrara, A., Pisante, M., Quaranta, G., & Taberner, M. (2000). Evaluating environmental sensitivity at the basin scale through the use of geographic information systems and remotely sensed data: an example covering the Agri basin - Southern Italy. *Catena*, *40*, 19-35.

Biasi, R., Brunori, E., Ferrara, C. & Salvati, L. (2017). Towards sustainable rural landscapes? a multivariate analysis of the structure of traditional tree cropping systems along a human pressure gradient in a Mediterranean region. *Agroforestry Systems*, *91*(6), 1199-1217.

Biasi, R., Colantoni, A., Ferrara, C., Ranalli, F. & Salvati, L. (2015). In-between sprawl and fires: Long-term forest expansion and settlement dynamics at the wildland-urban interface in Rome, Italy. *International Journal of Sustainable Development and World Ecology*, *22*(6), 467-475.

Blaikie, P. & Brookfield, H. C. (1987). *Land degradation and society*. Methuen, London.

Bonavero, P., Dematteis, G. & Sforzi, F. (1999). *The Italian Urban System*. Towards European Integration. Ashgate, Aldershot.

Brandt, J. (2005). *Desertification information system to support National Action Programmes in the Mediterranean (DISMED)*. DIS4ME, Desertification Indicator System for Mediterranean Europe (www. unibas.it/desertnet/dis4me/using-dis4me /dismed.htm downloaded in March 2008).

Briassoulis, H. (2005). *Policy integration for complex environmental problems*. Ashgate, Aldershot.

Carlucci, M., Grigoriadis, E., Rontos, K. & Salvati, L. (2017). Revisiting a Hegemonic Concept: Long-term 'Mediterranean Urbanization' in Between City Re-polarization and Metropolitan Decline. *Applied Spatial Analysis and Policy*, *10*(3), 347-362.

Ceccarelli, T., Giordano, F., Luise, A., Perini, L. & Salvati, L. (2006). *Vulnerability to desertification in Italy: collection, analysis, comparison, and validation of procedures for risk mapping and indicators used at national, regional and local scale.* National Agency for Environmental Protection, Rome - Technical Report No. 40 (in Italian).

Ceccarelli, T., Bajocco, S., Salvati, L. & Perini, L. (2014). Investigating syndromes of agricultural land degradation through past trajectories and future scenarios. *Soil Science and Plant Nutrition*, *60*(1), 60-70.

Chopra, K. & Gulati, S. C. (1997). Environmental degradation and population movements: the role of property rights. *Environmental and Resource Economics*, *9*, 383-408.

Cimini, D., Tomao, A., Mattioli, W., Barbati, A. & Corona, P. (2013). Assessing impact of forest cover change dynamics on high nature value farmland in Mediterranean mountain landscape. *Annals of Silvicultural Research*, *37*(1), 29-37.

Colantoni, A., Mavrakis, A., Sorgi, T. & Salvati, L. (2015a). Towards a 'polycentric' landscape? Reconnecting fragments into an integrated network of coastal forests in Rome. *Rendiconti Lincei*, *26*, 615-624.

Colantoni, A., Ferrara, C., Perini, L. & Salvati, L. (2015b). Assessing trends in climate aridity and vulnerability to soil degradation in Italy. *Ecological Indicators*, *48*, 599-604.

Colantoni, A., Grigoriadis, E., Sateriano, A., Venanzoni, G. & Salvati, L. (2016). Cities as selective land predators? A lesson on urban growth, deregulated planning and sprawl containment. *Science of the Total Environment*, 545-546, 329-339.

Conacher, A. J. (2000). *Land degradation.* Kluwer Academic Publishers, Dordrecht.

Conacher, A. J. & Sala, M. (1998). *Land degradation in Mediterranean environments of the world.* Wiley, Chichester.

Corona, P., Fattorini, L. & Franceschi, S. (2009). Estimating the volume of forest growing stock using auxiliary information derived from relascope or ocular assessments. *Forest Ecology and Management*, *257*(10), 2108-2114.

Corona, P., Ascoli, D., Barbati, A., Bovio, G., Colangelo, G., Elia, M. & Lovreglio, R. (2014). Integrated forest management to prevent wildfires under mediterranean environments. *Annals of Silvicultural Research*, *38*(2), 24-45.

Cuadrado-Ciuraneta, S., Durà-Guimerà, A. & Salvati, L. (2017). Not only tourism: unravelling suburbanization, second-home expansion and “rural” sprawl in Catalonia, Spain. *Urban Geography*, *38*(1), 66-89.

D’Angelo, M., Enne, G., Madrau, S., Percich, L., Previtali, F., Pulina, G. & Zucca, C. (2000). Mitigating land degradation in Mediterranean agro-silvo-pastoral systems: a GIS-based approach. *Catena*, *40*, 37-49.

De Rosa, S. & Salvati, L. (2016). Beyond a ‘side street story’? Naples from spontaneous centrality to entropic polycentricism, towards a ‘crisis city’. *Cities*, *51*, 74-83.

Delfanti, L., Colantoni, A., Recanatesi, F., Bencardino, M., Sateriano, A., Zambon, I. & Salvati, L. (2016). Solar plants, environmental degradation and local socioeconomic contexts: A case study in a Mediterranean country. *Environmental Impact Assessment Review*, *61*, 88-93.

Di Feliciantonio, C. & Salvati, L. (2015). 'Southern' Alternatives of Urban Diffusion: Investigating Settlement Characteristics and Socio-Economic Patterns in Three Mediterranean Regions. *Tijdschrift voor Economische en Sociale Geografie*, *106*(4), 453-470.

Dore, M. H. I. (2005). Climate Change and Changes in Global Precipitation Patterns: What Do We Know? *Environment International*, *31*(8), 1167-1181.

Duvernoy, I., Zambon, I., Sateriano, A. & Salvati, L. (2018). Pictures from the other side of the fringe: Urban growth and peri-urban agriculture in a post-industrial city (Toulouse, France). *Journal of Rural Studies*, *57*, 25-35.

Garcia Latorre, J., Garcia-Latorre, J. & Sanchez-Picon, A. (2001). Dealing with aridity: socio-economic structures and environmental changes in an arid Mediterranean region. *Land Use Policy*, *18*, 53-64.

Geist, H. J. & Lambin, E. F. (2004). Dynamic causal patterns of desertification. *Bioscience*, *54*, 817-829.

Harte, J. (2007). Human population as a dynamic factor in environmental degradation. *Population and Environment*, *28*, 223-236.

Kairis, O., Karavitis, C., Kounalaki, A., Salvati, L. & Kosmas, C. (2013). The effect of land management practices on soil erosion and land desertification in an olive grove. *Soil Use and Management*, *29*(4), 597-606.

Kairis, O., Karavitis, C., Salvati, L., Kounalaki, A. & Kosmas, K. (2015). Exploring the impact of overgrazing on soil erosion and land degradation in a dry Mediterranean agro-forest landscape (Crete, Greece). *Arid land research and management*, *29*(3), 360-374.

Karamesouti, M., Detsis, V., Kounalaki, A., Vasiliou, P., Salvati, L. & Kosmas, C. (2015). Land-use and land degradation processes affecting soil resources: Evidence from a traditional Mediterranean cropland (Greece). *Catena*, *132*, 45-55.

Kazemzadeh-Zow, A., Zanganeh Shahraki, S., Salvati, L. & Samani, N. N. (2017). A spatial zoning approach to calibrate and validate urban growth models. *International Journal of Geographical Information Science*, *31*(4), 763-782.

Kelly, C., Ferrara, A., Wilson, G. A., Ripullone, F., Nolè, A., Harmer, N. & Salvati, L. (2015). Community resilience and land degradation in forest and shrubland socio-ecological systems: Evidence from Gorgoglione, Basilicata, Italy. *Land Use Policy*, *46*, 11-20.

Kok, K., Rothman, D. S. & Patel, M. (2004). Multi-scale narratives from an IA perspective: Part I. European and Mediterranean scenario development. *Futures*, *38*, 261-284.

Kosmas, C., Karamesouti, M., Kounalaki, K., Detsis, V., Vassiliou, P. & Salvati, L. (2016). Land degradation and long-term changes in agro-pastoral systems: An empirical analysis of ecological resilience in Asteroussia-Crete (Greece). *Catena*, *147*, 196-204.

Lambin, E. (1993). Spatial scales and desertification. *Desertification Control Bulletin*, *23*, 20-23.

Lawn, P. A. (2003). A theoretical foundation to support the Index of Sustainable Economic Welfare (ISEW), Genuine Progress Indicator (GPI), and other related indexes. *Ecological Economics*, *44*, 105-118.

Lemon, M., Seaton, R. & Park, J. (1994). Social enquiry and the measurement of natural phenomena: the degradation of irrigation water in the Argolid Plain, Greece. *International Journal of Sustainable Development and World Ecology*, *2*(3), 1-11.

Marchetti, M., Vizzarri, M., Lasserre, B., Sallustio, L. & Tavone, A. (2015). Natural capital and bioeconomy: challenges and opportunities for forestry. *Annals of Silvicultural Research*, *38*(2), 62-73.

Montanarella, L. (2007). *Trends in land degradation in Europe*. In: Sivakumar M. V., N'diangui, N. (Eds), Climate and land degradation. Springer, Berlin.

Munafò, M., Salvati, L. & Zitti, M. (2013). Estimating soil sealing rate at national level - Italy as a case study. *Ecological Indicators*, *26*, 137-140.

Oxley, T. & Lemon, M. (2003). From social-enquiry to decision support tools: towards an integrative method in the Mediterranean rural environment. *Journal of Arid Environment*, *54*, 595-617.

Pili, S., Grigoriadis, E., Carlucci, M., Clemente, M. & Salvati, L. (2017). Towards sustainable growth? A multi-criteria assessment of (changing) urban forms. *Ecological Indicators*, *76*, 71-80.

Puigdefabregas, J. & Mendizabal, T. (1998). Perspectives on desertification: western Mediterranean. *Journal of Arid Environment*, *39*, 209-224.

Recanatesi, F., Clemente, M., Grigoriadis, E., Ranalli, F., Zitti, M. & Salvati, L. (2016). A fifty-year sustainability assessment of Italian agro-forest districts. *Sustainability (Switzerland)*, *8*(1), 1-13.

Rontos, K., Grigoriadis, E., Sateriano, A., Syrmali, M., Vavouras, I. & Salvati, L. (2016). Lost in protest, found in segregation: Divided cities in the light of the 2015 "Οχι" referendum in Greece. *City*, *Culture and Society*, *7*(3), 139-148.

Salvati, L. (2013). Monitoring high-quality soil consumption driven by urban pressure in a growing city (Rome, Italy). *Cities*, *31*, 349-356.

Salvati, L. (2014). Agro-forest landscape and the 'fringe' city: A multivariate assessment of land-use changes in a sprawling region and implications for planning. *Science of the Total Environment*, *490*, 715-723.

Salvati, L. & Carlucci, M. (2011). The economic and environmental performances of rural districts in Italy: Are competitiveness and sustainability compatible targets? *Ecological Economics*, *70*, 12, 2446-2453.

Salvati, L. & Carlucci, M. (2014). A composite index of sustainable development at the local scale: Italy as a case study. *Ecological Indicators*, *43*, 162-171.

Salvati, L. & Zitti, M. (2005). Land degradation in the Mediterranean Basin: Linking bio-physical and economic factors into an ecological perspective. *Biota*, *6*(43132), 67-77.

Salvati, L. & Zitti, M. (2009). Assessing the impact of ecological and economic factors on land degradation vulnerability through multiway analysis. *Ecological Indicators*, *9*(2), 357-363.

Salvati, L., Petitta, M., Ceccarelli, T., Perini, L., Di Battista, F. & Scarascia, M. E. V. (2008a). Italy's renewable water resources as estimated on the basis of the monthly water balance. *Irrigation and Drainage*, *57*(5), 507-515.

Salvati, L., Zitti, M. & Ceccarelli, T. (2008b). Integrating economic and environmental indicators in the assessment of desertification risk: A case study. *Applied Ecology and Environmental Research*, *6*(1), 129-138.

Salvati, L., Zitti, M., Ceccarelli, T. & Perini, L. (2009). Developing a synthetic index of land vulnerability to drought and desertification. *Geographical Research*, *47*(3), 280-291.

Salvati, L., Bajocco, S., Ceccarelli, T., Zitti, M. & Perini, L. (2011). Towards a process-based evaluation of land vulnerability to soil degradation in Italy. *Ecological Indicators*, *11*(5), 1216-1227.

Salvati, L., Gemmiti, R. & Perini, L. (2012a). Land degradation in Mediterranean urban areas: An unexplored link with planning? *Area*, *44*(3), 317-325.

Salvati, L., Perini, L., Sabbi, A. & Bajocco, S. (2012b). Climate Aridity and Land-use Changes: A Regional-Scale Analysis. *Geographical Research*, *50*(2), 193-203.

Salvati, L., Morelli, V. G., Rontos, K. & Sabbi, A. (2013a). Latent exurban development: City expansion along the rural-to-urban gradient in growing and declining regions of southern Europe. *Urban Geography*, *34*(3), 376-394.

Salvati, L., Tombolini, I., Perini, L. & Ferrara, A. (2013b). Landscape changes and environmental quality: The evolution of land vulnerability and potential resilience to degradation in Italy. *Regional Environmental Change*, *13*(6), 1223-1233.

Salvati, L., Zitti, M. & Sateriano, A. (2013c). Changes in city vertical profile as an indicator of sprawl: Evidence from a Mediterranean urban region. *Habitat International*, *38*, 119-125.

Savo, V., De Zuliani, E., Salvati, L., Perini, L. & Caneva, G. (2012). Long-term changes in precipitation and temperature patterns and their possible impacts on vegetation (Tolfa-Cerite area, central Italy). *Applied Ecology and Environmental Research*, *10*(3), 243-266.

Smiraglia, D., Ceccarelli, T., Bajocco, S., Salvati, L. & Perini, L. (2016). Linking trajectories of land change, land degradation processes and ecosystem services. *Environmental Research*, *147*, 590-600.

Tanrivermis, H. (2003). Agricultural land-use change and sustainable use of land resources in the Mediterranean region of Turkey. *Journal of Arid Environment*, *54*, 553-564.

Tomao, A., Quatrini, V., Corona, P., Ferrara, A., Lafortezza, R. & Salvati, L. (2017). Resilient landscapes in Mediterranean urban areas: Understanding factors influencing forest trends. *Environmental Research*, *156*, 1-9.

Trisorio, A. (2005). *Measuring sustainability. Indicators for Italian agriculture.* National Institute of Agricultural Economics, Rome.

Trisorio-Liuzzi, G. & Hamdy, A. (2002). Desertification: causes and strategies to compete. *Options Méditerranéennes* (A), *50*, 399-412.

Vogt, J. V., Safriel, U., Bastin, G., Zougmore, R., von Maltitz, G., Sokona, Y. & Hill, J. (2011). Monitoring and Assessment of Land Degradation and Desertification: Towards new conceptual and integrated approaches. *Land Degradation and Development*, *22*(2), 150-165.

Wilson, G. A. & Juntti, M. (2005). *Unravelling desertification: policies and actor networks in Southern Europe*. Wageningen, Wageningen Academic Publishers.

Zambon, I., Serra, P., Sauri, D., Carlucci, M. & Salvati, L. (2017). Beyond the 'mediterranean city': Socioeconomic disparities and urban sprawl in three Southern European cities. *Geografiska Annaler, Series B: Human Geography*, *99*(3), 319-337.

Zambon, I., Benedetti, A., Ferrara, C. & Salvati, L. (2018). Soil Matters? A Multivariate Analysis of Socioeconomic Constraints to Urban Expansion in Mediterranean Europe. *Ecological Economics*, *146*, 173-183.

Zitti, M., Ferrara, C., Perini, L., Carlucci, M. & Salvati, L. (2015). Long-term urban growth and land-use efficiency in Southern Europe: Implications for sustainable land management. *Sustainability (Switzerland)*, *7*(3), 3359-3385.

Zuindeau, B. (2007). Territorial equity and sustainable development. *Environmental Values*, *16*, 253- 268.

In: Land Degradation: The Main Challenge ISBN: 978-1-53615-575-4
Editors: Ilaria Zambon et al.

Chapter 2

LAND DEGRADATION: A SOCIOECOLOGICAL PERSPECTIVE

***Luca Salvati**[1,*]**, Kostas Rontos**[2,†]**, Rosanna Salvia**[3,‡]*
***and Ilaria Zambon**[4,#]*
[1]Council for Agricultural Research and Economics (CREA), Arezzo, Italy
[2]University of the Aegean, Mytilene, Greece
[3]University of Basilicata, Potenza, Italy
[4]Tuscia University, Viterbo, Italy

ABSTRACT

Land degradation raises interesting research issues, when defining related processes in their conceptual and quantitative aspects, and when studying implications on natural resources and ecosystems. Understanding land degradation requires a multidisciplinary approach, in view of the different processes involved, as well as the acquisition of a considerable amount of basic information. Land degradation can evolve

[*] Corresponding Author's E-mail: luca.salvati@crea.gov.it
[†] Author's E-mail: k.rontos@soc.aegean.gr
[‡] Author's E-mail: rosanna.salvia@unibas.it
[#] Author's E-mail: ilaria.zambon@unitus.it

into an irreversible phase of 'desertification'. This term, although bringing the root 'desert' back to the foreground, should not be understood as a 'generator of deserts'; in many countries geographically far from desertic areas, it is possible to find latent conditions of desertification. Land degradation becomes manifest only when it is too late to withdraw from the irreversible conditions that were generated (Salvati et al., 2008a, 2009). According to United Nations data sources, a worrying picture emerges: 70% of arid arable land, corresponding to about 30% of total land, is affected by land degradation (Salvati et al., 2008b). Given the serious effects on populations and environments, the problem is particularly relevant in Africa, Asia, South America and the Caribbean. However, countries with structurally strong economies (e.g., United States, Australia, Japan and Europe) are also affected by land degradation, whose genesis and evolution, although attributable to a patchwork of different causes, appears unequivocally connected and reinforced by climate change.

2.1. Setting the Framework

Desert areas are part of Earth ecosystems and represent, like any other ecosystem, the equilibrium point of adaptive natural processes. The evolutionary dynamics of deserts is characterized by sequential phases of expansion and decline that evidence, from time to time, areas with unstable ecosystem balance. These transition zones represent the archetype of environmental marginality and are less resilient to anthropogenic pressures. Any area that experiences environmental imbalances and is subject to unsustainable anthropogenic pressure can potentially be affected by land degradation (Tanrivermis, 2003; Zuindeau, 2007; Salvati et al., 2008, 2009; Salvati and Carlucci, 2011, 2014; Zitti et al., 2015; Recanatesi et al., 2016; Pili et al., 2017). These processes represent a prototype of environmental degradation (Delfanti et al., 2016) and assume reversible or irreversible forms based on the factors involved, the resistance of the environmental matrix and the resilience of biotic and abiotic components (Briassoulis, 2005; Chelleri et al., 2015; Kosmas et al., 2016). Various conditions can aggravate (e.g., climate change and human activities) or mitigate (e.g., soil and vegetation factors) land degradation (Basso et al., 2000; Bajocco et al., 2015; Biasi et al., 2017). Processes of degradation include over-exploitation of aquifers for irrigation purposes: an excessive drainage, in addition to exhausting water table, causes a lowering of the groundwater level and the intrusion of the wedge of sea water into subsoil (Salvati et al., 2008a). Use of brackish water in turn causes salinization of the fertile layers of the soil, altering biotic and physical properties and determining a loss of organic matter (D'Angelo et al., 2000).

Conceptual frameworks interpreting land degradation processes had a long and complex evolution. These have inspired actions aimed at solving current problems rather than long-term mechanisms of desertification. The problem of desertification came to the fore in the 1930s, when the Great Plains of the United States of America became a dust bowl following a serious and persistent drought adding to massive exploitation of fertile agricultural land. Hundreds of thousand people were forced to leave their homeland and livelihoods to emigrate elsewhere, causing a serious socioeconomic crisis (Salvati et al., 2016). Subsequently, the adoption of more appropriate cultivation methods preserving water resources contributed to limit the negative consequences of such extreme events (Salvati et al., 2008a, 2009). Since the 1970s, the scientific debate has progressively focused on the concept of desertification, adopting definitions that are increasingly appropriate to describe causes and consequences of land degradation (Salvati el al., 2016). The United Nations Convention to Combat Drought and Desertification (UNCCD) states: 'Desertification is the degradation of land in arid, semi-arid and dry sub-humid areas, attributable to various causes, including climatic variations and human activities'. This definition identifies areas where desertification may occur, in turn indicating natural and anthropogenic causes of land degradation.

2.1.1. Operational Definitions of Land Degradation

Despite the recognized importance of a standardized definition to identify and share theoretical concepts and analytic approaches, the scientific debate on desertification was often characterized by abuse of general expressions raising confusion in stakeholders. In this case, likely because of its evocative power, the 'desertification' term represents an exemplary case (Salvati et al., 2012a). It cannot be excluded that lexical disorder may also arise from the myriad of potential causes of desertification, as well as from scaling factors. As the field of analysis narrows, in both spatial and temporal dimensions, the inherent simplification of the cause-effect links also leads to the simplification of the involved actors. For instance, while in some steep lands the notion of 'desertification' may resemble to water erosion, in other contexts, where agricultural activity is no longer profitable, the concept of 'desertification' is associated with 'land abandonment'. To recognize the relevant terminology, attempts have been made to identify environmental notions (Le Houerou, 1993). Land degradation was therefore defined as the (reversible) phenomenon of environmental degradation that manifests itself mainly through the reduction of wooded and/or cultivated areas accompanied by soil erosion (Delfanti et al., 2016). Conversely, the term 'desert' refers to a land degradation of human origin such as those resulting from intense migration from the countryside to urban and industrial areas. Finally, a widespread definition identifies 'desertification' as a degradation typical of transitional areas at the fringe of desertic contexts.

The operational definition of land degradation is flanked by terms widely used in scientific literature, such as 'sensitivity', 'vulnerability' and 'risk', which, however, are always used in ambiguous ways or used as synonyms (Salvati et al., 2008a, 2009). The need for clarity and unambiguousness of the desertification definition is particularly important in international contexts where correct communication is key to science and policy. In this sense, the IPCC (Intergovernmental Panel on Climate Change) clarified the meanings associated with 'sensitivity' and 'vulnerability' concepts. The first term should be used to indicate the 'level of degradation reached in a certain territory by processes of desertification due to climate change, soil erosion, deforestation, salinization, triggered by natural or human causes'; the second term instead expresses the 'level of susceptibility (or resilience) to desertification'.

In accordance with these definitions, the concept of land sensitivity to degradation is linked to the intensity of biophysical processes intended as an evolutionary consequence of past conditions, while the notion of vulnerability mainly refers to the future evolution of such processes. Land vulnerability is a function of intensity and severity of desertification and intrinsic sensitivity (and adaptive capability) of the associated socioecological system. According to the European Environment Agency (EEA) glossary, a sensitive area is as 'an area where special measures have to be taken to protect natural habitats with a high level of vulnerability'; vulnerability is the 'degree to which a system is susceptible, or unable, to cope with damage'; and the definition of risk focuses on the concept of 'expected losses (life, health, property, income) due to a particular hazard occurring in a certain area and period'.

In the context of the Desertlinks and DISforME Projects financed by the European Community, 'vulnerable area' is defined as a context where environmental, socioeconomic and land management factors are not balanced. In this case, 'environmental vulnerability' is seen as the result of interactions between various factors relating to soil, climate, vegetation and socioeconomic conditions (Salvati et al., 2017). For instance, the combination of critical factors (rough morphology, soils exposed to strong erosion, unfavorable weather conditions, poor vegetation cover), together with sub-optimal socioeconomic factors (intensive agriculture, mass tourism, soil sealing), identifies a high level of vulnerability (Zambon et al., 2017). For the United Nations Convention to Combat Desertification (UNCCD), the concept of 'vulnerability' is linked to the extent of damage caused by a change in the environmental balance in relation to the sensitivity of the system and to its ability to adapt to new conditions. In other contexts, 'environmental sensitivity' has been defined as the degree of responsiveness to ecological stresses produced by external anthropogenic and/or natural forces (Bajocco et al., 2012).

Finally, the Italian National Committee for the Fight against Desertification (CNLD) has defined vulnerable an 'area at risk of functional loss of fertility' following processes of strong (and sometimes irreversible) soil degradation. In 'a vulnerable area', the quality of soils is compromised and close to conditions of functional sterility, despite resilience factors (e.g., vegetation cover, soil irrigation) able to mitigate desertification (Salvati and Zitti, 2012; Chelleri et al., 2015; Kosmas et al.,

2016). CNLD defined 'sensitive' land when processes leading to desertification, although active, have not yet led to the functional sterility of the soil (National Committee to Combat Desertification, 1998). This review of definitions highlights the complexity of desertification and the importance of considering multiple disciplinary perspectives (Salvati et al., 2016). Achieving a common language that can facilitate dialogue between scientists, decision-makers and stakeholders allows identification of more effective measures to combat land degradation (Briassoulis, 2005; Wilson and Juntti, 2005; Karamesouti et al., 2015).

2.2. Syndromes of Land Degradation

Land degradation reflects various causes classified according to natural and anthropogenic origins. An exhaustive taxonomy of causes of desertification has not yet been realized, despite the countless initiatives undertaken at both national and international level. Causes of land degradation can be distinguished as biophysical or socioeconomic (Salvati and Zitti, 2005; Zambon et al., 2017; Biasi et al., 2017). The former includes climate change and soil degradation; the latter includes landscape transformations, population pressure and tourism. Theory should be accompanied by empirical observation, since degradation factors, and the resulting consequences, are not the linear result of a simple sum. In fact, they reflect complex interactions and determine cascading effects that make it complex to trace back to the primary causes of land syndromes (Galeotti, 2007). Causal chains due to the integrated action of several factors include: (i) depopulation of rural areas, increase of undergrowth vegetation, wildfires, landslides, (ii) urban expansion (especially in coastal areas), land-use changes, development of infrastructure, determining soil compaction and sealing (Munafò et al., 2013, Morelli et al., 2014), over-exploitation of water-table (intrusion of the saline wedge deeper inland, saline water abstraction for crop irrigation, soil sterility due to salinization). More recently, an innovative approach has emerged that is able to grasp complexity of the connection between interacting factors.

Reference is made to the concept of 'environmental syndrome' (Hill et al., 2008) which considers desertification as a causal chain in which various factors participate in a synergic way. The examples given above can be described through this new interpretative key.

2.2.1. Climate Change

Among the most relevant biophysical factors in desertification processes (Salvati and Zitti, 2005), climate change plays a key role. The term 'climatic factor' is used here to refer to climatic characteristics in the strict sense, but also to atmospheric phenomena attributable to meteorological variability (Colantoni et al., 2015a). Climate can directly promote soil degradation due to the beating of rain, the intensity of wind or late frosts. Climatic factors can also make an indirect contribution to soil degradation (Salvati, 2014a), for example playing a key role in the geographical distribution of plant species that provide soil cover and protection (Mendelsohn and Dinar, 2003; Salvati et al., 2012a, b; Recanatesi et al., 2013). In addition to climate factors, there are also elements of change linked to the so-called 'global warming'. Beyond anthropogenic causes, global warming represents a reality with which we are confronted at local level. Given the specificity of territories potentially exposed to desertification, the consequences of global warming should be measured as environmental, social and cultural changes (Incerti et al., 2007).

One of the most tangible consequences of global warming is changes in rainfall regime, leading climate aridity (Dore, 2005). Aridity is a structural characteristic of local climates, due to the combination of low rainfall (< 400 mm/year) and intense evapotranspiration rate, which create soil conditions unsuitable for vegetation. To assess the degree of aridity, the UNCCD has adopted an index calculated by the ratio of the annual rainfall to the potential average annual evapotranspiration rate. This index has been widely used for its simplicity and was originally proposed by UNEP (United Nation Environmental Program) in the World Atlas of

Desertification for the definition of lands at risk (Rubio et al., 2009). Although having the same effects as aridity, drought is a transitory phenomenon able to affect any territorial area in any season of the year (Salvati et al., 2009, 2012; Colantoni et al., 2015b). Drought occurs when the amount of rainfall is significantly lower than normal ('meteorological' drought). The resulting economic, environmental and social damages appear to be proportionally linked to the duration of drought (Le Houerou, 1993; Barbier, 2000). Time span determines the crucial element for assessing the degree of severity, the effects of which may progressively affect wider social and economic sectors. In this regard, a drought is defined as 'agricultural', thanks to its intensity and duration, pushing the primary sector to crisis; a drought is defined as 'hydrological', if it can create an emergency for the industrial sector and civil uses by affecting the flow of water bodies. Natural ecosystems, being the result of gradual adaptation to environmental conditions, generally possess an intrinsic resilience that allows overcoming the 'physiological' periods of drought. Conversely, anthropogenic agro-ecosystems can suffer damage of varying intensity due to the interference generated by human activities on resilience mechanisms (Chelleri et al., 2015; Kosmas et al., 2016; Recanatesi et al., 2016). Drought can also affect areas with arid climates where the use of water resources is close to the critical threshold of environmental sustainability (Salvati et al., 2009).

Climate change caused by rainfall patterns is also manifested through the different distribution of precipitation along the year. The frequency of extreme events (heavy rain) has increased and rainfall is concentrated in more restricted periods of the year (Salvati et al., 2008). This can lead to an increase in the erosiveness of rainfall, where sloping soils or land without vegetation has also led to a significant increase in hydrogeological instability (Salvati et al., 2009). Degradation processes resulting from atmospheric phenomena can also be emphasized by latent factors that increase their effects: highly erodible substrates (calcareous rocks or clayey-sandy sedimentary formations), soil fragility (shallow soil layer, lack of structure, low organic matter content, low permeability), peculiar land morphologies (soil exposure and gradient), water use (excessive or unplanned abstraction, point and diffuse pollution from different sources), and quality of vegetation cover (Ferrara et al., 2014, 2017). Continuity and variety of vegetation cover are essential for soil protection (Bajocco et al., 2012; Salvati and Zitti, 2012; Kairis et al., 2013; Colantoni et al., 2015b; Kosmas et al., 2016; Fares et al., 2017).

2.2.2. Soil Degradation

In accordance with what has been mentioned above, land degradation processes are particularly evident where conditions of climatic vulnerability and fragility of ecosystems mainly exist (Salvati and Carlucci, 2012). As recognized by the Organization for Economic Co-operation and Development (OECD), soil degradation processes can be categorized into physical, chemical and biological factors according to the nature of the proximate causes. The former includes those which result in soil loss (volume and surface subtraction). These forms of degradation are, in many cases, the result of a land management that has not been able to combine principles of economic and productive development with the needs of conservation of soil resources (Karamesouti et al., 2015). Among the most important processes of physical degradation, it is possible to mention: (i) soil erosion, due to the action of water and wind that physically remove soil particles (sometimes erosion is also triggered by human activities such as intensive farming techniques practiced in hilly areas, deforestation and wildfires); (ii) soil compaction, due to compression of soil particles and consequent reduction of its porosity (in this case, the cause is excessive mechanical pressure due to heavy machinery or the excessive grazing); (iii) soil sealing, mainly caused by construction of buildings and roads, which removes large areas of soils from their natural functions, altering the balance of the ecosystem (production of biomass, absorption and filtration of rainwater). The second class of degradation processes includes environmental phenomena that cause deterioration of the soil chemical characteristics (Salvati, 2014b; Kairis et al., 2013, 2015; Kosmas et al., 2016). These include contamination (particularly in industrial areas, mining areas and major transport routes) and salinization. The latter can be distinguished into primary or secondary salinization if it is caused by characteristics already in the geo-lithology of the place or produced by external contributions of salt substances (e.g., using irrigation salt water). Finally, biological degradation affects soil functionality (fertility, resistance to erosion, buffering capacity) and biodiversity (Bajocco et al., 2011; Canfora et al., 2017; Renzi et al., 2017).

2.2.3. The Social Context

Anthropogenic pressure on ecosystems often represents an unsustainable exploitation of natural resources. This alters the environmental balance leading to a reduction of natural capital, which is progressively replaced by economic capital. The main causes of human pressure are due to agriculture, animal husbandry, over-exploitation of water resources, wildfires, urbanization, industrialization and concentration of seasonal tourism (Cuadrado-Ciuraneta et al., 2017; Zambon et al., 2018). The growth of tourism becomes an element of strong impact when combined with other factors of pressure and when exerted on ecologically-fragile areas (Thornes and Brandt, 1996; Mairota et al., 1998; Conacher, 2000). For instance, tourism pressure in coastal areas has exerted a significant impact on local communities in relation to different settlement patterns (Di Feliciantonio and Salvati, 2015; Cuadrado-Ciuraneta et al., 2017) leading to a shortage of services such as water supply and urban waste disposal (Venezian Scarascia et al., 2006; Salvati et al., 2008a). In the long term, the causal chain triggered by tourism pressure can lead to a significant loss of agricultural land. Previous studies show conflicting views on the link between tourism development and land degradation and identify potentially beneficial effects (Makhzoumi, 1997; Loumou et al., 2000; Iosifides and Politidis, 2005; Onate and Peco, 2005). These effects can be observed at various spatial and temporal scales (Figure 1). At regional level, tourism development can stimulate greater environmental awareness and protection through policy responses (Briassoulis, 2004), or through increased awareness of the natural heritage that has had positive effects on the local economy. However, a holistic approach based on qualitative and quantitative assessment is needed to identify optimal development strategies for governing environments and the related socioeconomic contexts (Wilson and Juntti, 2005).

Figure 1. Traditional village in Mani, Greece (left); urban sprawl and soil consumption in a tourism place in Messinia, Greece (right). Source: own elaboration.

Anthropogenic pressures caused by agriculture, depend on technically incorrect practices, implemented with the aim of maximizing quantitative yields in the short-term through unsustainable use of the means of production, such as excessive use of chemical fertilizers and pesticides, excessive use of water resources, lack of (unsuitable) crop rotation (Biasi et al., 2015a). In such cases, the most evident consequences are soil constipation, superficial crusting, loss of chemical-physical fertility, progressive salinization of the surface soil layers and pollution of the aquifers. Land abandonment, a phenomenon that affects economically-marginal agricultural systems, is also the starting point for soil degradation, up to the extreme consequences of hydrogeological instability. Livestock activities, if carried out without considering environmental characteristics, can lead to soil impoverishment and/or contamination triggering land degradation processes. Although in recent years there has been an overall reduction in the number of livestock in the Mediterranean region, the environmental impact of livestock farming has remained unchanged or has increased due to high specialized animal husbandry. In many cases, livestock has been introduced in unsuitable areas, causing serious imbalances in agro-ecosystems, above all, for the substantial pollution of the surface watercourses and aquifers by nitrates.

Even in marginal uplands of hills and mountains, where suitable areas are restricted, the livestock load per unit of surface area can become unsustainable as the activity is concentrated above all where there is easier access to pastures and water, road networks and availability of electricity (Salvati et al., 2013a; Recanantesi et al., 2016). In traditional pastures, problems arise from overgrazing, which leads to the destruction of grassland and to the consequent exposure of soils to water and/or wind erosion. Moreover, the effects of some 'traditional' practices, such as wildfires in wooded areas to create new pastures, should not be overlooked (Salvati and Ferrrara, 2014). In addition to the physical and chemical properties of the soil erosion, wildfires affect the 'waterproofing' of eroded soil, making it more exposed to erosion processes. It should not be forgotten that wildfires alter ecosystems by changing composition and structure of plant and animal biocoenoses. Although ecosystems have the capacity to recover over long periods of time, wildfires compromise natural responses to desertification. It should also be noted that a main cause of wildfires in the Mediterranean region was illegal urbanization on fringe land (Biasi et al., 2015a; De Rosa et al., 2016; Duvernoy et al., 2018), having clear implications on soil sealing (Munafò et al., 2013) and alteration of hydrological regimes on a local scale (Salvati et al., 2008). In recent decades, the demand of water for residential, agricultural and industrial uses has led to a sharp increase in water abstractions and derivations, compromising the renewal processes of the resource. Agriculture, again, is the main process since it accounts for about 60-70% of total annual water consumption. The progressive expansion of intensive crops into traditional cultivation systems, has stimulated an increase in irrigation requirements, especially under climate change.

While agriculture is the activity that absorbs most of water resources, the importance of water withdrawals for residential use cannot be overlooked, particularly in areas of high population density e.g., coastal areas, large urban areas, tourism sites. In fact, the unsustainable use of water resources has led, especially in recent decades, to a sharp increase in water consumption. Urban expansion, in addition to water consumption, involves the removal of significant amounts of fertile soils, both for direct sealing of land surfaces and for land-use changes driven by population growth (roads, shopping malls, landfills). These phenomena are also correlated with various processes of soil contamination, which lead to more complex forms of soil degradation, as well as the consequent fragmentation of rural landscapes (Munafò et al., 2013).

2.3. Anthropogenic Causes of Land Degradation

The above arguments form a basic knowledge for developing more sophisticated schemes that allow identification and classification of socioeconomic factors leading to land degradation (Zambon et al., 2017). This scheme is particularly relevant in the operationalization of the above-mentioned concepts through specific indicators. In this regard, 6 types of variables are considered (Figure 2):

- location (time and space) and severity of land degradation, considered as a dependent variable;
- the actors involved (individuals, families, companies or institutions that influence the vulnerability of a given territory to land degradation);
- decisions determining the level of land degradation;
- the external variables influencing the choices of the actors involved;
- macroeconomic variables (i.e., factor that influence land vulnerability through decision-making variables);
- policy tools, i.e., variables able to contrast or mitigate land degradation.

In this context, the role of decisions taken by specific actors should be highlighted. In fact, the preferences of territorial actors, influenced by boundary conditions (e.g., technology, institutions, services, market and available infrastructures), have a decisive role on the choices made and, therefore, become themselves a cause of land degradation (Briassoulis, 2011). However, a clear identification of the interacting factors is necessary to highlight the individual variables according to the interpretative model (e.g., microeconomic approaches focused on immediate causes or macroeconomic models dealing with causes). Whatever the approach, it suffers from the limitations of current knowledge of process dynamics and the causal links between the individual variables involved (Wilson and Juntti, 2005).

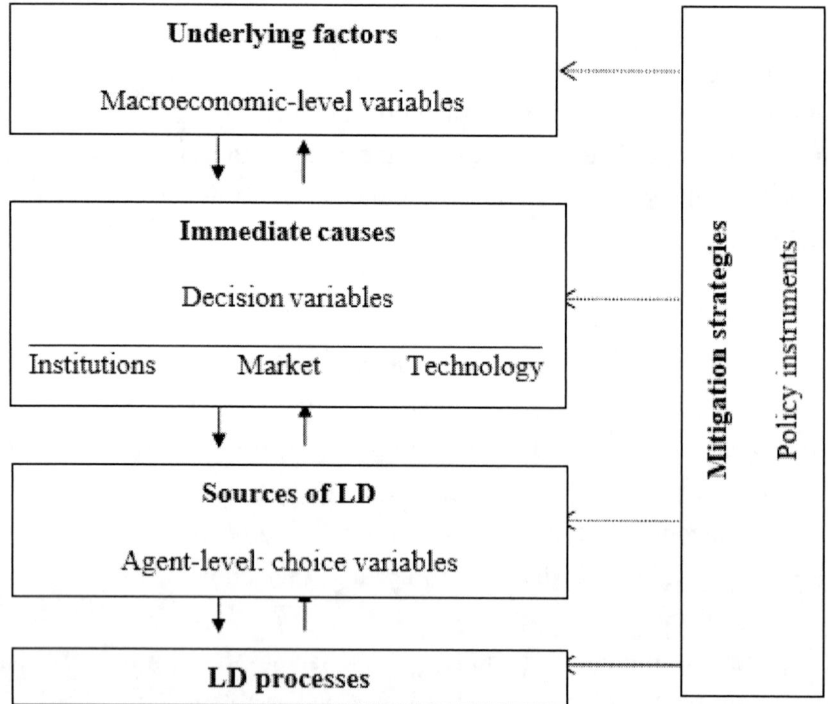

Figure 2. Sources, immediate causes, underlying factors, strategies against desertification, and their relations in the Mediterranean region. Source: own elaboration.

2.3.1. Immediate Causes

Unsustainable practices, particularly in the primary sector, appear to be one of the main determinants of land degradation in Mediterranean rural areas (Mendelsohn and Dinar, 2003). Changes in agricultural technology, prices of agricultural products and inputs, ownership regimes and soil quality are the most relevant causes of land degradation (Biasi et al., 2015b). Technology has a direct impact on the behavior of agricultural operators and an indirect impact resulting from its impact on product and factor prices (Salvati and Carlucci et al., 2011). Technological changes that tend to increase production without altering labor intensity, can have a local impact on land degradation *via* crop intensification (Pender, 1998; Rubio et al., 2009; Salvati, 2010). The impact could be even more significant if technological changes tend to reduce the labor force or capital employed, as this frees up resources that can be used to further intensify farming practices (Abdelgalil and Cohen, 2001). In these circumstances, however, the net effect on land degradation is indirect in nature and difficult to assess (Pender et al., 2004). This evidence suggests that scientific research and agricultural policies should propose technologies that are more profitable. However, empirical evidence is still limited, and this topic is an important objective for future studies (Coxhead and Jayasuriya, 1994; Briassoulis, 2005; Salvati and Carlucci et al., 2011).

The increased profitability of some agricultural products may lead to over-exploitation of land while exacerbating soil degradation. Conversely, the increased availability of capital because of higher incomes may recover more marginal or previously abandoned farmland. This is not the case when farmers show preferences for one type of subsistence agriculture, but this logic seems to apply only to a limited extent in rural areas of southern Europe. Changes in the cost of production can also trigger negative soil phenomena linked to the adoption of unsustainable cropping systems (Blaikie and Brookfield, 1987). Without well-defined property rights, there are various reasons why land can be degraded, and this is particularly true in poor countries in Africa, South America and East Asia (Beaumont and Walker, 1996; Chopra and Gulati, 1997; Jayasuriya, 2003; Danfeng et al., 2006). In this perspective, studies in Mediterranean Europe are quite limited. Theoretically, two mechanisms should be considered: first, although profits may be negative in the early years, technological and infrastructure progress can make crops profitable in the future and push farmers to occupy new land. Second, in many cases land prices do not directly reflect soil quality, being indicative of speculative aspects of selling land for urban use (Mairota et al., 1998; Hubacek and van der Bergh, 2006; Karamesouti et al., 2015). This aspect can be observed, for example, in areas situated in the outskirts of Mediterranean cities (Figure 3).

However, it should be noted that the above processes could play a limited role in the Mediterranean region where ownership regimes are generally well defined in time and space. However, the relationship between land rent and land management should be further explored in the light of soil degradation, e.g., in lowland areas devoted to agriculture and experiencing soil erosion and salinization (Marathianou et al., 2000; Lemon et al., 1994; Perez-Sirvent et al., 2003; Mendelsohn and Dinar, 2003; Salvati et al., 2012; Recanatesi et al., 2013).

Figure 3. Transition from agricultural land into open (abandoned) spaces converted to built-up areas at the fringe of Mediterranean cities: Athens, Greece (left) and Rome, Italy (right). Source: own elaboration.

2.3.2. Latent Factors

It is quite difficult to identify the socioeconomic factors underlying land degradation and to establish meaningful links between them (Zambon et al., 2017). This is particularly true for macroeconomic variables that influence environmental conditions through complex pathways and often indirect relationships (Galeotti, 2007). Studies investigating such issues often face incomplete or poor-quality information and data. Only disaggregated data allow the possibility of developing models capable of analyzing complex territorial systems. These limits are constraints to the development of studies aimed at exploring jointly the spatial-temporal dimensions of land degradation (Salvati and Zitti, 2005÷ Salvati, 2014b; Zambon et al., 2018). Nevertheless, in such cases it is possible to describe qualitatively some of the possible latent factors underlying land degradation, including economic growth, population expansion and the consequent anthropogenic pressure (Zitti et al., 2015; Salvati et al., 2016; Kazemzadeh-Zow et al., 2017).

Human pressure has been intensified over the last 50 years (Harte, 2007; Biasi et al., 2017). A growing population triggers needs and expectations of quality of life that lead to urbanization, population concentration, migration, increasingly intense land-use determining a positive correlation between population density and level of land degradation (Karamesouti et al., 2015). A basic assumption regarding the relationship between land degradation and population dynamics indicates that migration could lead to a discrepancy between economic carrying capacity and population density in dry or arid regions affected by land degradation (Figure 4). Carrying capacity is the ability of the economic system to sustain a defined population, e.g., the maximum number of people who can live in each region, based on the physical and economic resources to maintain a standard of living in the long term. The carrying capacity of a given region is thus determined by its physical capital, human resources, the technology used, as well as institutional arrangements, and the possibility to exchange goods and services with the outside world. Such dynamics, although not necessarily linked to land degradation, create potentially unstable situations that are detrimental to the natural environment (Salvati and Zitti, 2008).

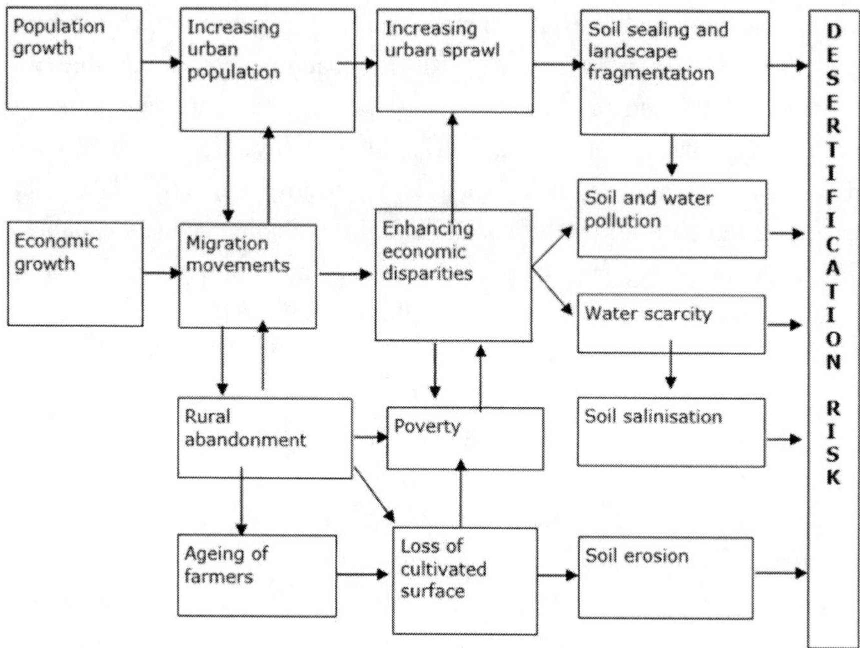

Figure 4. Demographic and economic factors affecting land resource depletion and increasing desertification risk. Source: own elaboration.

In several cases, especially in combination with other factors, it has been demonstrated that population pressure can act as a catalyst for land degradation (Cuffaro, 2001). High population increase does not necessarily lead to land degradation (Ferrara et al., 2017). What is more important is the mix of (i) sensitivity and fragility of land, (ii) the rate of population increase, and (iii) other critical forces, e.g., social and economic conditions, land-use and settlements patterns and cultivation practices (Genske, 2003; Kelly et al., 2015). Therefore, a causal connection between population growth and land degradation or a stable 'carrying capacity' of land beyond of which the problem starts to worsen seems to be no simply determined. The absence of a direct connection between population pressure and land degradation, must not lead to the demise of this factor as a primary cause of environmental degradation. Combining biophysical and socioeconomic forces (Salvati and Zitti, 2005; Zambon et al., 2017), population pressure may be the catalyst for the intensification and severity of land degradation (Salvati, 2014). This phenomenon becomes potentially harmful, especially for inland areas, because of the negative effects associated with rural depopulation (Salvati et al., 2008). Equally, due to increased economic availability, food demand could increase, leading to intensification and depletion of the most fertile agricultural land (Figure 5). However, the cumulative effects of these processes have not been sufficiently explored up to date. What emerges from empirical analysis suggests the existence of a 'U' relationship that occurs between the indicators of environmental quality and the level of per capita income (Salvati and Carlucci, 2014). In the recent past, Environmental Kuznets Curve (EKC) has sought to address the relationship between environmental degradation and economic growth (Figure 6). Recent studies have demonstrated the validity of such a framework even if the hypothesis remains controversial and often criticized (Galeotti, 2007). The EKC hypothesis is valid for some environmental processes affecting local areas, while its validity at larger scales is less tangible (Arrow et al., 1995; Grossman and Krueger, 1995; McConnell, 1997). The theoretical basis for verifying the existence of an EKC on the depletion of natural resources, particularly soil, is complex because of the difficulty of measuring land vulnerability in relation to its predominant causes (Stern, 2004).

Figure 5. Historical settlements in Mani, Greece (left) and Basilicata, Italy. Source: own elaboration.

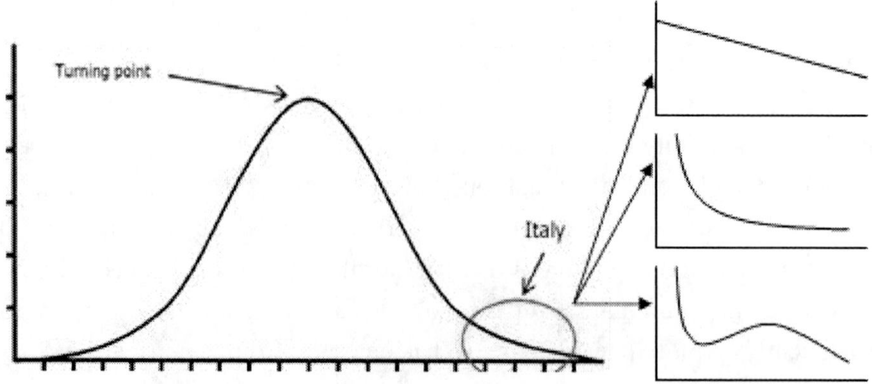

Figure 6. Possible forms of the global relationship among land degradation (x-axis) and income (y-axis) according to the EKC hypothesis and the relative position of Italy as representative of Southern European countries. Source: own elaboration.

Technological developments appear to be a key element determining economic growth (Galeotti, 2007). An example is given by those technologies that improve market access increasing, on the one hand, the consumption of natural resources to adapt production to new demand and encouraging the reduction of human pressure on the environment for delocalization of production (Wilson and Juntti, 2005; Salvati, 2013, 2014). However, based on current knowledge, it seems difficult to propose a definitive scheme at macroscopic scale as further investigations are necessary.

Finally, development policies can also have indirect effects on land degradation. It is well known, for example, that agricultural policies that focus on subsidies to support the production of individual crops can stimulate the conversion of traditional, multi-functional and sustainable farming systems into intensive systems that are often not adapted to local environmental conditions (Juntti and Wilson, 2004). Such options are due to the failure to consider the costs and benefits deriving from them, as happened with the subsidization of mono-cultural production systems that are not compatible with the typical conditions of the Mediterranean basin. The costs of converting production systems should consider all the factors involved (including environmental factors) to analyze correctly the cost-benefit ratio. Earlier studies have demonstrated that the costs of conversion of traditional agricultural systems into intensive agricultural systems (especially cereal systems) are considerably higher than the economic benefits obtainable when the costs of negative externalities (erosion, salinization, compaction, water scarcity and fragmentation of the landscape) are also considered (Briassoulis, 2004).

2.4. Economic Consequences of Land Degradation

If it is true that socioeconomic factors play an important role in determining the level of land degradation of a certain territory (Zambon et al., 2017), it is also true that land degradation has socioeconomic consequences triggering a spiral from which it is difficult to emerge without adequate policies (Scherr, 2000; Briassoulis, 2011). In the most extreme conditions of degradation, social, political and institutional conflicts can arise because of the scarcity of natural resources (Briassoulis, 2005). More generally, land degradation has a negative impact on crop productivity; low land productivity in arid districts, coupled with uncertain yields, discourage investment and technological innovation (Le Houerou, 1993; Barbier, 2000), leading to a regressive land management and weak economic performance, further exacerbating soil degradation (Pender, 1998; Pender et al., 2004). Distortions that result from these factors increase disparities within the rural sector, e.g., between rich and poor areas, generating potentially disruptive dynamics and pressures on local populations.

In these regards, land degradation leads to depopulation of rural areas, starting from marginal districts, with a consequent relocation of population to urban and coastal areas. Abandoned agricultural land will tend towards denaturalization or desertification to a level that depends on the state of the land and the ecological factors involved (Kosmas et al., 2000; Atis, 2006; Hein, 2007; Ibanez et al., 2008; Salvati and Zitti, 2009). In Europe, since the end of the World War II, there has been a generalized abandonment of land, which has affected mainly the countries of the Mediterranean basin because of industrialization processes, low agricultural incomes and rigid markets (Le Houerou, 1993).

The relationship between ecosystem carrying capacity, population density and land degradation is controversial in many ways; in conditions of aridity and low soil fertility, population growth (due to birth and/or immigration) can lead to generalized poverty and can accelerate disparities between social classes (Zuindeau, 2007). Migration is a complex phenomenon which often relates, directly or indirectly, to various problems of environmental degradation. Migration can be analyzed in several ways, among which the temporal dimension and the direction of movement are important. About the temporal dimension, a distinction is made between seasonal, semi-permanent and permanent movements. Depending on the direction, migrations can occur, either within the national territory or between different countries.

Poverty, understood as a general condition of the population, can provide a key to explain migration and the consequent abandonment of land (rural areas) and/or demographic concentration (urban areas) with the consequent expansion of fringes. In southern Europe, the geography of social impoverishment often coincides with arid environments and, indirectly, with structural conditions of vulnerability to desertification. Also, in this case, causal mechanisms of land degradation are not explored and complex factors concerning the organization of society, such as political structure, governance strategies, interventions of market regulation, are involved in such processes (Iosifides and Politidis, 2005). When the political strategy appears inadequate to cope with environmental problems and the market and/or institutions are weak, poverty can accelerate land degradation and *vice versa*. Migratory movements generated by economic crisis lead to additional conditions of poverty due to the difficulty, ensuring that the entire population has adequate employment and wages. Once again, this process particularly affects the weakest social actors, such as women, young people and currently also the over-50s, indirectly contributing to other social problems with a local and regional impact.

All the processes analyzed so far contribute to economic polarization by consolidating and/or amplifying social inequalities between rich and disadvantaged areas (Rontos et al., 2016; Carlucci et al., 2017; Duvernoy et al., 2018; Cuadrado-Ciuraneta et al., 2017). National and local strategies aimed at mitigating socioeconomic imbalances should reflect the tight of the 2007 Lisbon Treaty, giving priority to the need of implementing sustainable economic growth. Correcting these imbalances may benefit from a linkage between socioeconomic issues and mitigation of environmental pressure framework (Salvati, 2016).

2.5. Analysis of Land Degradation Processes

Assessment of land degradation makes use of analysis' tools capable of grasping multifaced aspects using the basic information available. Over the years, as knowledge and awareness were consolidating, the approach to analyze land degradation has been enriched with more sophisticated tools. This evolution can be deduced from a comparison of interpretative schemes, representing criteria and relationships considered in the early 2000s and the more recent years, which reflect the progress of monitoring techniques, as well as the crucial role of socioeconomic determinants (Basso et al., 2000; Briassoulis, 2011). Applying the DPSIR (Drivers-Pressures-State-Impact-Responses) scheme to the study of land degradation implies identification of drivers, pressure, state, impact and responses. Starting from a reference hypothesis, the following degradation systems have been identified (Montanarella, 2007): climate and climate change, soil erosion, urbanization, soil salinization, soil pollution, compaction of soil surface layers, soil degradation due to forest fires and other anthropogenic impacts (Mendelsohn and Dinar, 2003; Salvati et al., 2012; Cimini et al., 2013; Recanatesi et al., 2013; Corona et al., 2014; Biasi et al., 2015a; Colantoni et al., 2015b; Marchetti et al., 2015). Each of them interacts with the others, even if systems are analyzed separately, in accordance with the DPSIR approach which implies selection of multiple environmental indicators to describe land degradation.

The DPSIR scheme is in line with previous studies on desertification in Mediterranean environments. For instance, Camci Cetin et al. (2007), indicate two complex processes as the main causes of vulnerability to desertification in selected Mediterranean contexts, these processes can be easily identified in the DPSIR scheme described above as follows:

- the increasing human pressure in coastal and flat areas because of urban growth, internal migration, concentration of tourism and intensive agriculture, in turn linked to unsustainable use of water resources and soil salinization, compaction, sealing and diffuse soil contamination (Loumou et al., 2000; Tanrivermis, 2003; Iosifides and Politidis, 2005; Munafò et al., 2013; Zitti et al., 2015; Salvati et al., 2016; Cuadrado-Ciuraneta et al., 2017; Kazemzadeh-Zow et al., 2017; Duvernoy et al., 2018);
- depopulation of marginal and mountainous agricultural areas, with the consequent abandonment of land in connection with the increase in heavy rainfall and forest fires, which are causes of soil deterioration, especially in sloping areas with low vegetation cover (Garcia Latorre et al., 2001; Bajocco et al., 2012; Salvati and Zitti, 2012; Biasi et al., 2015a; Colantoni et al., 2015a).

References

Abdelgalil, E.A., and Cohen, S.I. (2001). Policy modelling of the trade-off between agricultural development and land degradation - the Sudan case. *J. Policy Model.*, 23, 847-874.

Arrow, K., Bolin, B., Costanza, R., Dasgupta, P., Folke, C., Holling, C.S., Jansson, B-O., Levin, S., Mäler, K-G., Perrings, C., and Pimentel, D. (1995). Economic Growth, Carrying Capacity, and the Environment. *Science,* 268, 520-521.

Atis, E. (2006). Economic impacts on cotton production due to land degradation in the Gediz Delta, Turkey. *Land use policy*, 23, 181-186.

Bajocco, S., De Angelis, A., and Salvati, L. (2012). A satellite-based green index as a proxy for vegetation cover quality in a Mediterranean region. *Ecological Indicators*, 23, 578-587.

Bajocco, S., Dragozi, E., Gitas, I., Smiraglia, D., Salvati, L., and Ricotta, C. (2015). Mapping forest fuels through vegetation phenology: The role of coarse-resolution satellite time-series. *PLoS ONE*, 10(3), e0126430.

Bajocco, S., Salvati, L., and Ricotta, C. (2011). Land degradation versus fire: A spiral process? *Progress in Physical Geography*, 35(1), 3-18.

Barbier, E.B. (2000). The economic linkages between rural poverty and land degradation: some evidence from Africa. *Agriculture, Ecosystems and Environment*, 82(1-3), 355-370.

Basso, F., Bove, E., Dumontet, S., Ferrara, A., Pisante, M., Quaranta, G., and Taberner, M. (2000). Evaluating environmental sensitivity at the basin scale through the use of geographic information systems and remotely sensed data: an example covering the Agri basin - Southern Italy. *Catena*, 40, 19-35.

Beaumont, P.M., and Walker, R.T. (1996). Land degradation and property regimes. *Ecological Economics*, 18, 55-66.

Biasi, R., Brunori, E., Ferrara, C., and Salvati, L. (2017). Towards sustainable rural landscapes? a multivariate analysis of the structure of traditional tree cropping systems along a human pressure gradient in a Mediterranean region. *Agroforestry Systems*, 91(6), 1199-1217.

Biasi, R., Colantoni, A., Ferrara, C., Ranalli, F., and Salvati, L. (2015a). In-between sprawl and fires: Long-term forest expansion and settlement dynamics at the wildland-urban interface in Rome, Italy. *International Journal of Sustainable Development and World Ecology*, 22(6), 467-475.

Biasi, R., Brunori, E., Smiraglia, D., and Salvati, L. (2015b). Linking traditional tree-crop landscapes and agro-biodiversity in central Italy using a database of typical and traditional products: a multiple risk assessment through a data mining analysis. *Biodiversity and Conservation*, 24(12), 3009-3031.

Blaikie, P., and Brookfield, H.C. (1987). *Land degradation and society*. Methuen, London.

Briassoulis, H. (2004). The institutional complexity of environmental policy and planning problems: the example of Mediterranean desertification. *Journal of Environmental Planning and Management* 47, 115-135.

Briassoulis, H. (2005). *Policy integration for complex environmental problems*. Ashgate, Aldershot.

Briassoulis, H. (2011). Governing desertification in Mediterranean Europe: the challenge of environmental policy integration in multi-level governance contexts. *Land Degradation and Development*, 22(3), 313-3.

Camci Cetin, S., Karaca, A., Haktanir, K., and Yand ildiz, H. (2007). Global attention to Turkey due to desertification. *Environmental Monitoring and Assessment,* 128, 489-493.

Canfora, L., Salvati, L., Benedetti, A., and Francaviglia, R. (2017). Is soil microbial diversity affected by soil and groundwater salinity? Evidences from a coastal system in central Italy. *Environmental Monitoring and Assessment*, 189(7), 319.

Carlucci, M., Grigoriadis, E., Rontos, K., and Salvati, L. (2017). Revisiting a Hegemonic Concept: Long-term 'Mediterranean Urbanization' in Between City Re-polarization and Metropolitan Decline. *Applied Spatial Analysis and Policy,* 10(3), 347-362.

Chelleri, L., Schuetze, T., and Salvati, L. (2015). Integrating resilience with urban sustainability in neglected neighborhoods: Challenges and opportunities of transitioning to decentralized water management in Mexico City. *Habitat International,* 48, 122-130.

Chopra, K., and Gulati, S.C. (1997). Environmental degradation and population movements: the role of property rights. *Environmental and Resource Economics,* 9, 383-408.

Cimini, D., Tomao, A., Mattioli, W., Barbati, A., and Corona, P. (2013). Assessing impact of forest cover change dynamics on high nature value farmland in Mediterranean mountain landscape. *Annals of Silvicultural Research*, 37(1), 29-37.

Colantoni, A., Ferrara, C., Perini, L., and Salvati, L. (2015b). Assessing trends in climate aridity and vulnerability to soil degradation in Italy. *Ecological Indicators,* 48, 599-604.

Colantoni, A., Mavrakis, A., Sorgi, T., and Salvati, L. (2015a). Towards a 'polycentric' landscape? Reconnecting fragments into an integrated network of coastal forests in Rome. *Rendiconti Lincei*, 26, 615-624.

Conacher, A.J. (2000). *Land degradation.* Kluwer Academic Publishers, Dordrecht.

Corona, P., Ascoli, D., Barbati, A., Bovio, G., Colangelo, G., Elia, M., and Lovreglio, R. (2014). Integrated forest management to prevent wildfires under mediterranean environments. *Annals of Silvicultural Research*, 38(2), 24-45.

Coxhead, I., and Jayasuriya, S. (1994). Technical Change in Agriculture and Land Degradation in Developing Countries: A General Equilibrium Analysis. *Land Economics,* 70(1), 20-37.

Cuadrado-Ciuraneta, S., Durà-Guimerà, A., and Salvati, L. (2017). Not only tourism: unravelling suburbanization, second-home expansion and 'rural' sprawl in Catalonia, Spain. *Urban Geography*, 38(1), 66-89.

Cuffaro, N. (2001). *Population, economic growth and agriculture in less developed countries.* Routledge, New York, USA.

D'Angelo, M., Enne, G., Madrau, S., Percich, L., Previtali, F., Pulina, G., and Zucca, C. (2000). Mitigating land degradation in Mediterranean agro-silvo-pastoral systems: a GIS-based approach. *Catena,* 40, 37-49.

Danfeng, S., Dawson, R., and Baoguo, L. (2006). Agricultural causes of desertification risk in Minquin, China. *Journal of Environmental Management,* 79, 348-356.

De Rosa, S., and Salvati, L. (2016). Beyond a 'side street story'? Naples from spontaneous centrality to entropic polycentricism, towards a 'crisis city'. *Cities,* 51, 74-83.

Delfanti, L., Colantoni, A., Recanatesi, F., Bencardino, M., Sateriano, A., Zambon, I., and Salvati, L. (2016). Solar plants, environmental degradation and local socioeconomic contexts: A case study in a

Mediterranean country. *Environmental Impact Assessment Review*, 61, 88-93.

Di Feliciantonio, C., and Salvati, L. (2015). 'Southern' Alternatives of Urban Diffusion: Investigating Settlement Characteristics and Socio-Economic Patterns in Three Mediterranean Regions. *Tijdschrift voor Economische en Sociale Geografie*, 106(4), 453-470.

Dore, M.H.I. (2005). Climate Change and Changes in Global Precipitation Patterns: What Do We Know? *Environment International,* 31(8), 1167-1181.

Duvernoy, I., Zambon, I., Sateriano, A., and Salvati, L. (2018). Pictures from the other side of the fringe: Urban growth and peri-urban agriculture in a post-industrial city (Toulouse, France). *Journal of Rural Studies*, 57, 25-35.

Fares, S., Bajocco, S., Salvati, L., Camarretta, N., Dupuy, J.-L., Xanthopoulos, G., Guijarro, M., Madrigal, J., Hernando, C., and Corona, P. (2017). Characterizing potential wildland fire fuel in live vegetation in the Mediterranean region. *Annals of Forest Science*, 74(1), 1.

Ferrara, A., Salvati, L., Sabbi, A., and Colantoni, A. (2014). Soil resources, land cover changes and rural areas: Towards a spatial mismatch? *Science of the Total Environment,* 478, 116-122.

Ferrara, C., Carlucci, M., Grigoriadis, E., Corona, P., and Salvati, L. (2017). A comprehensive insight into the geography of forest cover in Italy: Exploring the importance of socioeconomic local contexts. *Forest Policy and Economics*, 75, 12-22.

Galeotti, M. (2007). Economic growth and the quality of the environment: taking stock. *Environment, Development and Sustainability*, 9, 427-454.

Garcia Latorre, J., Garcia-Latorre, J., and Sanchez-Picon, A. (2001). Dealing with aridity: socio-economic structures and environmental changes in an arid Mediterranean region. *Land Use Policy*, 18, 53-64.

Genske, D.D. (2003). *Urban land - Degradation, investigation, remediation*. Springer, Berlin.

Grossman, G.M., and Krueger, A.B. (1995). Economic growth and the environment. *The Quarterly Journal of Economics*, 110, 353-377.

Harte, J. (2007). Human population as a dynamic factor in environmental degradation. *Population and Environment*, 28, 223-236.

Hein, L. (2007). Assessing the costs of land degradation: a case study for the Puentes catchment, southeast Spain. *Land Degradation and Development*, 18, 631-642.

Hill, J., Stellmes, M., Udelhoven Th., Roder A., and Sommer, S. (2008). Mediterranean desertification and land degradation: mapping related land-use change syndromes based on satellite observations. *Global and Planetary Change*, 64(3-4), 146-157.

Hubacek, K., and van den Bergh, J.C.J.M. (2006). Changing concepts of ‚land' in economic theory: from single to multi-disciplinary approaches. *Ecological Economics,* 56, 5-27.

Ibanez, J., Martinez Valderrama, J., and Puigdefabregas, J. (2008). Assessing desertification risk using system stability condition analysis. *Ecological Modelling*, 213, 180-190.

Incerti, G., Feoli, E., Salvati, L., Brunetti, A., and Giovacchini, A. (2007). Analysis of bioclimatic time series and their neural network-based classification to characterise drought risk patterns in South Italy. *International Journal of Biometeorology*, 51(4), 253-263.

Iosifides, T., and Politidis, T. (2005). Socioeconomic dynamics, local development and desertification in western Lesvos, Greece. *Local Environment,* 10, 487-499.

Jayasuriya, R.T. (2003). Measurement of the scarcity of soil in agriculture. *Resource Policy*, 29, 119-129.

Juntti, M., and Wilson, G.A. (2004). Conceptualising desertification in southern Europe: stakeholders' interpretations and multiple policy agendas. *European Environment,* 15, 228-249.

Kairis, O., Karavitis, C., Kounalaki, A., Salvati, L., and Kosmas, C. (2013). The effect of land management practices on soil erosion and land desertification in an olive grove. *Soil Use and Management*, 29(4), 597-606.

Kairis, O., Karavitis, C., Salvati, L., Kounalaki, A., and Kosmas, K. (2015). Exploring the impact of overgrazing on soil erosion and land degradation in a dry Mediterranean agro-forest landscape (Crete, Greece). *Arid Land Research and Management,* 29(3), 360-374.

Karamesouti, M., Detsis, V., Kounalaki, A., Vasiliou, P., Salvati, L., and Kosmas, C. (2015). Land-use and land degradation processes affecting soil resources: Evidence from a traditional Mediterranean cropland (Greece). *Catena,* 132, 45-55.

Kazemzadeh-Zow, A., Zanganeh Shahraki, S., Salvati, L., and Samani, N.N. (2017). A spatial zoning approach to calibrate and validate urban growth models. *International Journal of Geographical Information Science*, 31(4), 763-782.

Kelly, C., Ferrara, A., Wilson, G.A., Ripullone, F., Nolè, A., Harmer, N., and Salvati, L. (2015). Community resilience and land degradation in forest and shrubland socio-ecological systems: Evidence from Gorgoglione, Basilicata, Italy. *Land Use Policy*, 46, 11-20.

Kosmas, C., Gerontidis, S., and Marathianou, M. (2000). The effect of land-use changes on soil and vegetation over various lithological formations on Lesvos. *Catena*, 40, 51-68.

Kosmas, C., Karamesouti, M., Kounalaki, K., Detsis, V., Vassiliou, P., and Salvati L. (2016). Land degradation and long-term changes in agro-pastoral systems: An empirical analysis of ecological resilience in Asteroussia - Crete (Greece). *Catena*, 147, 196-204.

Kosmas, C., Tsara, M., Moustakas, N., and Karavitis, C. (2003). Identification of indicators for desertification. *Annals of Arid Zones*, 42, 393-416.

Le Houerou, H.N. (1993). Land degradation in Mediterranean Europe: can agroforestry be a part of the solution? A prospective review. *Agroforestry Systems*, 21, 43-61.

Lemon, M., Seaton, R., and Park, J. (1994). Social enquiry and the measurement of natural phenomena: the degradation of irrigation water in the Argolid Plain, Greece. *International Journal of Sustainable Development and World Ecology*, 2(3), 1-11.

Loumou, A., Giourga, C., Dimitrakopoulos, P., and Koukoulas, S. (2000). Tourism contribution to agro- ecosystems conservation; the case of Lesbos island, Greece. *Environmental Management*, 26, 363- 370.

Mairota, P., Thornes, J.B., and Geeson, N. (1998). *Atlas of Mediterranean environments in Europe.* The desertification context, Wiley, Chichester.

Makhzoumi, J.M. (1997). The changing role of rural landscapes: olive and carob multi-use tree plantations in the semiarid Mediterranean. *Landscape and Urban Planning*, 37, 115-122.

Marathianou, M., Kosmas, C., Gerontidis, S., and Detsis, V. (2000). Land-use evolution and degradation in Lesvos (Greece): a historical approach. *Land Degradation and Development,* 11, 63- 73.

Marchetti, M., Vizzarri, M., Lasserre, B., Sallustio, L., and Tavone, A. (2015). Natural capital and bioeconomy: challenges and opportunities for forestry. *Annals of Silvicultural Research*, 38(2), 62-73.

McConnell, K.E. (1997). Income and the demand for environmental quality. *Environment and Development Economics*, 2, 383-399.

Mendelsohn, R., and Dinar, A. (2003). Climate, water, and agriculture. *Land Economics*, 79, 328-341.

Montanarella, L. (2007). Trends in land degradation in Europe. In: Sivakumar M.V., N'diangui, N. (Eds.), *Climate and land degradation.* Springer, Berlin.

Morelli, V.G., Rontos, K., and Salvati, L. (2014). Between suburbanisation and re-urbanisation: revisiting the urban life cycle in a Mediterranean compact city. *Urban Research and Practice*, 7(1), 74-88.

Munafò, M., Salvati, L., and Zitti, M. (2013). Estimating soil sealing rate at national level - Italy as a case study. *Ecological Indicators,* 26, 137-140.

Onate, J.J., and Peco, B. (2005). Policy impact on desertification: stakeholders' perceptions in southeast Spain. *Land Use policy*, 22, 103-114.

Pender, J. (1998). Population growth, agricultural intensification, induced innovation and natural resource sustainability: an application of neoclassical growth theory. *Agricultural Economics*, 19, 99- 112.

Pender, J., Nkonya, E., Jagger, P., Sserunkuuma, D., and Ssali, H. (2004). Strategies to increase agricultural productivity and reduce land degradation: evidence from Uganda. *Agricultural Economics*, 31, 181-195.

Perez-Sirvent, C., Martinez-Sanchez, M.J., Vidal, J., and Sanchez, A. (2003). The role of low-quality irrigation water in the desertification of semi-arid zones in Murcia, SE Spain. *Geoderma*, 113, 109- 125.

Pili, S., Grigoriadis, E., Carlucci, M., Clemente, M., and Salvati, L. (2017). Towards sustainable growth? A multi-criteria assessment of (changing) urban forms. *Ecological Indicators*, 76, 71-80.

Recanatesi, F., Clemente, M., Grigoriadis, E., Ranalli, F., Zitti, M., and Salvati, L. (2016). A fifty-year sustainability assessment of Italian agro-forest districts. *Sustainability (Switzerland*), 8(1), 1-13.

Recanatesi, F., Ripa, M.N., Leone, A., Perini, L., and Salvati, L. (2013). Land-use, climate and transport of nutrients: Evidence emerging from the lake vicocase study. *Environmental Management*, 52(2), 503-513.

Renzi, G., Canfora, L., Salvati, L., and Benedetti, A. (2017). Validation of the soil Biological Fertility Index (BFI) using a multidimensional statistical approach: A country-scale exercise. *Catena,* 149, 294-299.

Rontos, K., Grigoriadis, E., Sateriano, A., Syrmali, M., Vavouras, I., and Salvati, L. (2016). Lost in protest, found in segregation: Divided cities in the light of the 2015 'Οχι' referendum in Greece. *City, Culture and Society*, 7(3), 139-148.

Rubio, J.L., and Bochet, E. (1998). Desertification indicators as diagnosis criteria for desertification risk assessment in Europe. *Journal of Arid Environment*, 39, 113-120.

Salvati, L. (2013). Monitoring high-quality soil consumption driven by urban pressure in a growing city (Rome, Italy*). Cities*, 31, 349-356.

Salvati, L. (2014a). Agro-forest landscape and the 'fringe' city: A multivariate assessment of land-use changes in a sprawling region and implications for planning. *Science of the Total Environment*, 490, 715-723.

Salvati, L. (2014b). The spatial pattern of soil sealing along the urban-rural gradient in a Mediterranean region. *Journal of Environmental Planning and Management*, 57(6), 848-861.

Salvati, L., and Carlucci, M. (2011). The economic and environmental performances of rural districts in Italy: Are competitiveness and sustainability compatible targets? *Ecological Economics,* 70(12), 2446-2453.

Salvati, L., and Carlucci, M. (2014). A composite index of sustainable development at the local scale: Italy as a case study. *Ecological Indicators*, 43, 162-171.

Salvati, L., and Ferrara, A. (2014). Do land cover changes shape sensitivity to forest fires in peri-urban areas? *Urban Forestry & Urban Greening*, 13(3), 571-575.

Salvati, L., and Zitti, M. (2005). Land degradation in the Mediterranean Basin: Linking bio-physical and economic factors into an ecological perspective. *Biota,* 6(43132), 67-77.

Salvati, L., and Zitti, M. (2008). Assessing the impact of ecological and economic factors on land degradation vulnerability through multiway analysis. *Ecological Indicators*, 9, 357-363.

Salvati, L., and Zitti, M. (2009). Substitutability and weighting of ecological and economic indicators: Exploring the importance of various components of a synthetic index. *Ecological Economics*, 68(4), 1093-1099.

Salvati, L., and Zitti, M. (2012). Monitoring vegetation and land-use quality along the rural-urban gradient in a Mediterranean region. *Applied Geography*, 32(2), 896-903.

Salvati, L., Petitta, M., Ceccarelli, T., Perini, L., Di Battista, F., and Venezian-Scarascia, M.E. (2008a). Italy's renewable water resources as estimated on the basis of the monthly water balance. *Irrigation and Drainage*, 57(5), 507-515.

Salvati, L., Zitti, M., and Ceccarelli, T. (2008b). Integrating economic and environmental indicators in the assessment of desertification risk: A case study. *Applied Ecology and Environmental Research*, 6(1), 129-138.

Salvati, L., Zitti, M., Ceccarelli, T., and Perini, L. (2009). Developing a synthetic index of land vulnerability to drought and desertification. *Geographical Research,* 47(3), 280-291.

Salvati, L., Gemmiti, R., and Perini, L. (2012a). Land degradation in Mediterranean urban areas: An unexplored link with planning? *Area*, 44(3), 317-325.

Salvati, L., Perini, L., Sabbi, A., and Bajocco, S. (2012b). Climate Aridity and Land-use Changes: A Regional-Scale Analysis. *Geographical Research*, 50(2), 193-203.

Salvati, L., Morelli, V.G., Rontos, K., and Sabbi, A. (2013a). Latent exurban development: City expansion along the rural-to-urban gradient in growing and declining regions of southern Europe. *Urban Geography*, 34(3), 376-394.

Salvati, L., Tombolini, I., Perini, L., and Ferrara, A. (2013b). Landscape changes and environmental quality: The evolution of land vulnerability and potential resilience to degradation in Italy. *Regional Environmental Change*, 13(6), 1223-1233.

Salvati, L., Sateriano, A., and Grigoriadis, E. (2016). Crisis and the city: profiling urban growth under economic expansion and stagnation. *Letters in Spatial and Resource Sciences*, 9(3), 329-342.

Scherr, S.J. (2000). A downward spiral? Research evidence on the relationship between poverty and natural resource degradation. *Food Policy,* 25, 479-498.

Stern, D.I. (2004). The rise and fall of the Environmental Kuznets Curve. *World Development*, 8, 1419- 1439.

Tanrivermis, H. (2003). Agricultural land-use change and sustainable use of land resources in the Mediterranean region of Turkey. *Journal of Arid Environment,* 54, 553-564.

Thornes, J.B., and Brandt, J. (1996). *Mediterranean desertification and land-use.* Chichester.

Venezian Scarascia, M.E., Di Battista, F., and Salvati, L. (2006). Water resources in Italy: availability and agricultural uses. *Irrigation and Drainage*, 55, 115-127.

Wilson, G.A., and Juntti, M. (2005). *Unravelling desertification: policies and actor networks in Southern Europe*. Wageningen, Wageningen Academic Publishers.

Zambon, I., Serra, P., Sauri, D., Carlucci, M., and Salvati, L. (2017). Beyond the 'mediterranean city': Socioeconomic disparities and urban sprawl in three Southern European cities. *Geografiska Annaler, Series B: Human Geography,* 99(3), 319-337.

Zambon, I., Benedetti, A., Ferrara, C., and Salvati, L. (2018). Soil Matters? A Multivariate Analysis of Socioeconomic Constraints to Urban Expansion in Mediterranean Europe. *Ecological Economics*, 146, 173-183.

Zitti, M., Ferrara, C., Perini, L., Carlucci, M., and Salvati, L. (2015). Long-term urban growth and land-use efficiency in Southern Europe: Implications for sustainable land management. *Sustainability* (Switzerland), 7(3), 3359-3385.

Zuindeau, B. (2007). Territorial equity and sustainable development, *Environmental Values*, 16, 253-268.

In: Land Degradation: The Main Challenge ISBN: 978-1-53615-575-4
Editors: Ilaria Zambon et al.

Chapter 3

SUSTAINABLE DEVELOPMENT AND POLICIES AGAINST DESERTIFICATION

Carlotta Ferrara[1,*], Massimo Cecchini[2,†], Luca Salvati[1,‡] and Ilaria Zambon[2,§]

[1]Council for Agricultural Research and Economics (CREA), Arezzo, Italy

[2]Tuscia University, Viterbo, Italy

ABSTRACT

During the last century, many countries experienced considerably (direct and indirect) impacts on environmental quality that are not yet fully understood. The relationship between economic growth and the consumption of natural capital makes it difficult to identify scenarios for a truly sustainable development (Salvati, 2013). For instance, the interactions between economic development, social inequalities and governance policies, in relation with environmental quality, should be better addressed (Galeotti, 2007; Salvati et al., 2013a; Cimini et al., 2013;

* Corresponding Author's Email: carlotta.ferrara@crea.gov.it

† Author's E-mail: cecchini@unitus.it

‡ Author's E-mail: luca.salvati@crea.gov.it

§ Author's E-mail: ilaria.zambon@unitus.it

Corona et al., 2014; Colantoni et al., 2015a; Marchetti et al., 2015; Rontos et al., 2016). In this respect, it is essential to identify processes that act as immediate causes and those representing latent factors of land degradation, as well as to recognize their effects in both the short- and long-term. The debate on sustainable development is focusing on complex interactions between bio-physical and socioeconomic factors, debating on the possibility of reconciling different concepts of 'sustainability' with the aim to formulate strategies promoting development and environmental quality together (Arrow et al., 1995; Smiraglia et al., 2016).

Land degradation is a clear example of what has been discussed so far in Mediterranean environments, where landscape and geographical characteristics are relatively fragile (Ibanez et al., 2008; Salvati et al., 2012a), human pressures have sometimes an irreversible impact on natural environments (Salvati and Zitti 2005, 2012; Bajocco et al., 2012; Salvati et al., 2012a, 2013a; Colantoni et al., 2015a; Di Feliciantonio and Salvati, 2015; Karamesouti et al., 2015; Delfanti et al., 2016; Zambon et al., 2017, 2018). In this perspective, the resulting 'net degradation' is mainly caused by anthropogenic factors (Mairota et al., 1998; Fernandez, 2002; Wilson and Juntti, 2005; Johnson and Lewis, 2007).

3.1. Sustainable Development and Land Desertification

The definition of 'sustainability' is historically linked to the Bruntland Report 'Our Common Future', drawn up in 1987 by the World Commission on Environment and Development (WCED), in which sustainable development is understood as a 'development that meets the needs of the present without compromising the ability of future generations to satisfy their own needs'. The concept of sustainability, directly linked to the economic flow of income between generations, thus extends to include the maintenance of quality and quantity of natural heritage, that is, the ensemble of goods whose existence, production and reproduction, although primarily due to the activity of nature, can be altered by human intervention. This definition, however, is quite general and the term 'future generations' is relatively vague. It is difficult to define the time horizon on which to calibrate the economic, political, moral and strategic choices to which entrust the task of guaranteeing development and the environment.

If, for example, companies and/or households are considered economic actors, the resulting time horizon would probably be too limited (two or three generations); for this reason, it might be appropriate to shift the focus from individual actors to organizations and institutions whose action, at least theoretically, requires longer time horizons to be effective. Further problems arise from the need to reconcile extreme positions such as (i) a degrowth strategy and (ii) business and economic development to preserve the environmental heritage. Between these two positions, there are studies that propose indicators combining development and wellbeing in the widest, most flexible and sophisticated way. It is not by chance, therefore, that per capita income is widely used as a variable capable of representing the degree and dynamics of development in a meaning that is not only strictly economic.

Sustainability also implies the possibility of substitution between different components of capital: thus, what is passed on to future generations is a generalized capacity to produce, rather than some specific components of capital. The latter interpretation responds to a so-called 'weak' rule of sustainability, according to which development is considered sustainable even if some components of capital decrease, provided that the total capital (natural and economic) does not decrease. However, scientific literature rejects the hypothesis of full (or partial) substitutability of natural capital (Arrow et al., 1995). According to this approach, which is more restrictive than the previous rule, the 'constant capital' rule prevails (i.e., 'strong' sustainability): natural capital should therefore remain constant (or increasing) under the assumption that capital stock remains constant or increases (Casadio Tarabusi and Palazzi, 2004). Balanced development is therefore the best compromise solution when considering all the components of the system (economic, social, environmental), with the objective of sustainability among its objectives. In the assessments of sustainable development, especially in a complex territorial system, it would be necessary to consider the spatial dimension to give it operational and concrete meaning to identify the so-called 'sustainable trajectory' of development (Zuindeau, 2007).

3.2. Ecological Economics

Depletion of natural resources and its consequences for biodiversity, climate change and soil degradation require definition of economic models (micro- and macro-level) that quantify the consumption of natural capital and assess its determinants and long-term impacts (e.g., Salvati et al., 2012b; Ceccarelli et al., 2014; Salvati, 2014; Biasi et al., 2015; Colantoni et al., 2015b; Zitti et al., 2015; Carlucci et al., 2017; Kazemzadeh-Zow et al., 2017). Structural changes in economic systems, together with historical and social factors and the system of public incentives, have controversial effects on soil degradation, with temporal and spatial dynamics not yet fully clarified. At a macro level, since the 1990s, there was considerable interest in the Environmental Kuznets Curve (EKC), which links the problem of environmental degradation to the economic well-being of countries, suggesting a non-linear relationship between the two variables (Galeotti, 2007). In the presence of low levels of economic development, the intensity of environmental degradation is limited to a minimum impact of subsistence activities and a low amount of waste in circulation. At a time when development is increasing - both through the intensification of agriculture, the take-off of the industrial sector or the increase in the rate of extraction of natural resources - the rate of regeneration of resources falls dramatically and the increase in waste, especially those with high toxicity, becomes significant. However, when development reaches extremely high levels, technological innovation, more effective regulation of production chains and increased public awareness of the environment can determine a gradual decline in environmental degradation.

The EKC is therefore represented by an inverted U-function whose maximum (usually called the turning point), represents the income level above which environmental degradation tends to be reduced in the presence of a simultaneous increase in income (Figure 1). The EKC report has been extensively used to test the relationship between income and air pollutants. Some applications have also evaluated the EKC model in deforestation processes. To date, an EKC approach to land degradation has not been developed, although the relationship between the two variables is necessarily controversial and influenced by ecological conditions rather than economic variables (Salvati et al., 2008a, 2008b; Salvati and Zitti, 2009; Salvati and Carlucci, 2011). Initially, the EKC models used linear econometric approaches to analyze groups of countries. This was done to highlight the existing dichotomy and the differential impact of income on the level of land degradation. Subsequently, the refinement of statistical techniques and the improved quality of basic data allowed the application of more sophisticated models (panel and time series), often supplemented by specific spatial analysis. More recently, some studies have shown the benefit of inter-country approaches rather than cross-country methodologies, particularly for those countries where income gaps and degradation levels are more marked and where the quality of basic information allows investigations even at an adequate geographical scale. It is evident, in this case, that the specific focus of the analysis shifts to the effectiveness of mechanisms of regulation/control, in a heterogeneous environmental context.

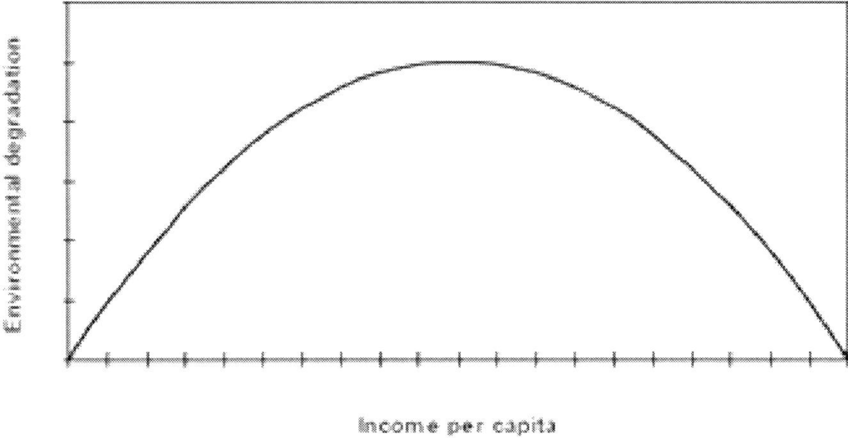

Figure 1. Illustration of a generic EKC relationship. Source: Own elaboration.

The EKC theory describes the above-mentioned relationship, by hypothesizing a development path in which, the increase of income elasticity according to environmental quality, is measured with changes in economic structure and different policy responses to explaining different development trajectories (Arrow et al., 1995). The main limits to the EKC hypothesis concern developing countries for aspects related to the simultaneity and influence of international trade, with reference to pollution-intensive goods (Galeotti, 2007), supporting the validity of the empirical findings in local short-term case studies.

Despite the Kuznets curve can be a valid logical reference and can represent an instrument of economic analysis in the context of a closer integration with the DPSIR framework, the effectiveness of responses to combat environmental degradation, can be evaluated precisely in relation to the income-degradation curve. In light of the EKC theory, a possible relationship between economic performance and land degradation is illustrated in Figure 2. The graph predicts that, in areas of higher economic well-being, negative externalities due to production processes can be mitigated by efficient environmental protection measures. It should be noted that the relationship tends to lose the typical quadratic trend that characterizes the EKC hypothesis, assuming instead, a linear form for high income values. Regional divides associated with spatially-varying problems of land degradation, suggest a more analytical approach than the EKC. In this case, more detailed information is needed to classify countries' land into relatively homogeneous systems.

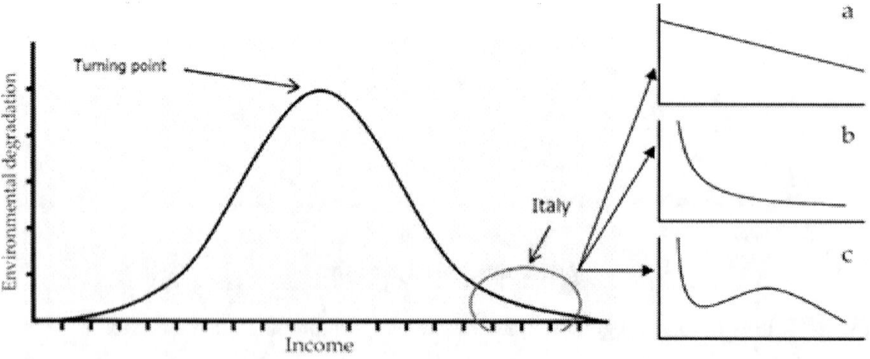

Figure 2. An EKC curve applied to land degradation in an advanced economy an example from Italy. Source: Own elaboration.

3.3. Environmental Sociology

From the late 1970s, social sciences began to provide a contribution to investigation of complex relationships between social and natural systems. Catton and Dunlap were responsible for '[...] the first attempt to elaborate a theoretical framework for a sociological analysis of the complex relationships that are established between society and the environment [...]' (Beato, 1998). To investigate such phenomena, social science should investigate the conditioning effects between biophysical contexts and social forces.

As far as the relationship with the natural environment is concerned, Western societies have displayed an anthropocentric attitude based on the optimistic prospect of ensuring unlimited progress and social development. It goes without saying that, in such a perspective, little importance has assumed the conditioning effect of the biophysical environment on the social systems and *vice versa*. Therefore, there was an urgent need to integrate the concept of limiting natural resources and thus their possible impoverishment into the cognitive approach of social sciences. This concept can be found in the theorization of the 'new ecological paradigm', proposed by Catton and Dunlap, which clearly shows the importance given to the concept of carrying capacity, i.e., the unavoidable link to the limits set by the natural environment and the non-negotiable cogency of ecological laws. The social structure, indeed, cannot be described and interpreted separately from its general context, since the claims-making of the social actors cannot be analyzed outside their social, historical-political and environmental references. Therefore, it is from the recognition of the inseparability between actors and social systems that the complex work of overcoming the dualism between Man and Nature begins.

The analytical structure proposed by Catton and Dunlap describes the action of society (considered in its constitutive elements: Population, Technology, Cultural System, Social System and Personality System) on the natural environment and, at the same time, the influence of the biophysical environment on the 'social complex'. The scheme elaborated by the two scholars is configured as a circular cause-effect structure (according to the sequence: social system → environment → social system), since it highlights '[...] a process of social causation of environmental change [...] whose effects fall on the structures and processes of society' (Beato, 1998). Moreover, contemporary societies are aware of the negative impact of their activities on natural systems (the so-called anthropogenic pressure on the environment). Following the emergence of the environmental crisis as a socio-political problem, '[...] they in turn develop 'responses' both regarding the structures and mechanisms of causation and the effects produced by environmental change [...]' such as, for instance, the implementation of environmental policies.

In this regard, Land (and Ecosystem) Degradation and Desertification (LEDD) are complex socio-environmental phenomena reflecting complex interaction of biophysical and societal forces on (and across) spatial scales. Land degradation may be changeable (within comparatively short time frames); if its drivers are left uncontrolled, land resources deteriorate leading to the irreparable state of desertification and to further reduction of ecosystem services. So, it gives rise to undesirable socioeconomic impacts. Recent analysis makes use of socio-environmental relationships to enrich the Complex Adaptive Systems (CAS) paradigm. The CAS paradigm predicts the inseparability and intertwined functioning of coupled socio-environmental organizations, the nonlinear relationships among their mechanisms and the existence of positive (and negative) feedback mechanisms. The task of designing effective responses to LEDD and of approachable policy making becomes the optimization of response assemblages (RAs), identifying those replies which are best to the features of LEDD in affected areas and which preserve the 'balance' among human activities (use) and available resources. Furthermore, an Optimal Response Assemblages (ORAs) comprise carefully synthesized, mutually helpful and coordinated responses, contributing to sustainable land management and societal welfare in affected regions (Kairis et al., 2013a; Mavrakis et al., 2015; Zitti et al., 2015; Delfanti et al., 2016).

3.4. Territorial Gaps In-Between Convergence and Polarizations

Many countries have experienced impressive economic development in recent decades, but regional disparities do not seem to have decreased significantly (Dinda, 2004; Stern, 2004; Galeotti, 2007; Carlucci et al., 2017). Disadvantaged in Portugal, Spain, Italy and Greece present critical conditions requiring integrated policy interventions. Per-capita income levels in these areas (many of which are predominantly rural) are still considerably lower than the European average. In Mediterranean countries, marked north-south dichotomy can be appropriately represented as a gradient of economic development. This gradient is consistent with the structural differentiation that can be seen regarding the level of economic well-being, which shows a significant decrease from north to south (Salvati et al., 2013b). For instance, analysis of the development paths of Italian regions shows that they differ significantly over the last fifty years. Northern Italy is regarded as a highly developed area from the economic and social point of view, while Southern Italy, despite a relative fast growth rate, still has a sharp economic gap testified by the predominance of the agricultural sector at the expense of industry. In Northern Italy, the industrial specialization has developed as an agglomerative response of diversification to stimuli typical of an industrial environment characterized by high economic heterogeneity. There is also a core of mature industries that has exploited productive dimensions promoting persistence of local employment models with high female and youth participation.

Research on regional convergence of economic, social and cultural variables provides the basic information to guide (sustainable) development policies (Barro and Sala-i-Martin, 2004). Convergence in environmental quality has been less analyzed than convergence in economic and social variables (Borzel, 2000; Neumayer, 2001; Jorgens, 2005; Aldy, 2006; Ezcurra, 2007). Applying a traditional convergence analysis for the level of vulnerability to land degradation in Italy, a complex convergence model was observed (Salvati et al., 2008a, 2011; Salvati and Carlucci, 2011, 2014; Munafò et al., 2013; Colantoni et al.,

2015b). Land vulnerability tends to diverge and polarize over time mainly in the regions of Northern and Central Italy (Ceccarelli et al., 2014), thus highlighting the influence of the geographical scale. This makes it important to differentiate between policies combatting desertification based on specific territorial features (Rubio et al., 2009; Kairis et al., 2013b, 2015; Kosmas et al., 2013). Two targets can be identified: (i) highly vulnerable land in disadvantaged economic conditions; (ii) moderately vulnerable land, with a divergent level of degradation over time, subject to climatic aridity and with a favorable socioeconomic context. Despite the acknowledged need to adopt policies mitigating degradation processes, only areas belonging to the first group have been considered so far, while, in the face of climate change and the global economic crisis, both target cases should be evaluated (Salvati et al., 2012, 2016; Colantoni et al., 2015b). Achieving an improved set of monitoring indicators is therefore a prerequisite for ensuring that policies to mitigate and reduce regional disparities are effective in all relevant cases (Kairis et al., 2013b).

3.5. Sustainability and Policy Strategies

Regional disparities may result from differences in geographic, historical, cultural and socioeconomic attributes. In the most disadvantaged areas, these symptoms are manifested in social backwardness, poor-quality education systems, high unemployment rates and low participation of women in the labor market, as well as partial or total inadequacy of infrastructure. European regional policy responds to these problems with instruments of financial solidarity that should act as a driving force for economic integration: more than one third of the Community budget is in fact transferred to the most disadvantaged regions. Sustainable growth appears to be one of the policy objectives of the European Community, envisioned in the Lisbon Strategy through the adoption of financial and institutional instruments (Pisani-Ferry, 2005; Ferrara et al., 2016; Pili et al., 2017). In brief, this strategy pursues ambitious objectives that are not restricted to the economic field alone (Tumpel-Gugerell and Mooslechner,

2003); promoting economic growth, innovation and employment, while preserving the inclusiveness of social models (Pisani-Ferry, 2005; Ceccarelli et al., 2014; Biasi et al., 2015; Zitti et al., 2015; Rontos et al., 2016; Carlucci et al., 2017; Kazemzadeh-Zow et al., 2017).

The assessment of the results achieved with the Lisbon Strategy, however, appears rather disappointing. The expected performance of the European economy in terms of growth, productivity and employment has generally not been achieved. Job creation has slowed down, while investments in research and development are relevantly low. The assessment of progress made by national governments under the Lisbon strategy is rather critical. The overloaded agenda, poor coordination and some missed priorities are highlighted. The Commission has therefore decided to focus on the actions to be taken rather than on the numerical targets to be achieved, updated as a re-launch of the political priorities for growth and jobs, but with a strong warning against the unsustainable use of natural resources.

Economic and social cohesion policies have been redesigned with the aim to fostering full integration between European, national and regional instruments to reach a high level of efficiency in achieving the sets of objectives (Zitti et al., 2015). The recent reform of the Structural Funds sets out several objectives that are considered crucial to economic growth, reduction of gender inequalities and fight against poverty. The reform was inspired by criteria of greater concentration of resources, decentralization and a stronger partnership. The 2007-2013 programming of the Structural Funds incorporated the contents of the reform based on the experience and results obtained in the previous wave and adopted methodological changes and new tools. A significant financial contribution is needed to achieve these objectives, e.g., from indirect Structural Funds, each focusing on a specific policy area:

- the European Regional Development Fund (ERDF), which operates in regions with a development deficit, undergoing economic conversion or facing structural difficulties;

- the European Social Fund (ESF), acting within the framework of the European Employment Strategy;
- the European Agricultural Guidance and Guarantee Fund (EAGGF), which finances rural development measures and provides support to farmers, particularly in disadvantaged regions under the Common Agricultural Policy (CAP).

3.6. Policies for Land Degradation

Territorial governance has assumed a significant role for land degradation (Blaikie and Brookfield, 1987). This is particularly true in developed countries, where civil society has become aware of the need to preserve environmental quality. Uncertainty and economic insecurity, which have emerged in recent years, may represent an obstacle to the effective application of environmental measures, especially in areas affected by climate aridity (Rubio et al., 2009). The impacts of environmental policies depend on a series of complex aspects that do not always allow an immediate and direct feedback on land, in light with the dynamic and non-linear nature of land degradation (Briassoulis, 2004; Thornes, 2004; Ibanez et al., 2008). The results obtained in several cases suggest that inaction (non-decision-making or *laissez-faire*) is a strategy with tangible effects on the environment and with costs not yet fully highlighted (Briassoulis, 2005).

There is a need for territorial policies to avoid fragmented approaches, promoting a full integration of short- and long-term strategies (Ceccarelli et al., 2014; Zitti et al., 2015; Carlucci et al., 2017), according to the spatial frame. In the case of specific measures developed for the agricultural sector (Duvernoy et al., 2018), the aim is usually to ensure income support and thus obtain or maintain a high degree of economic and social cohesion. Unfortunately, economic sustainability is not always combined with ecological sustainability (Mainguet, 1994; Steer, 1998; Conacher, 2000; Arshad and Martin, 2002) to the extent that, for instance, high-income intensive farming practices can be encouraged even where environmental resources would not allow it.

To offset the progressive reduction of agricultural income, national and European incentives have focused mainly on considering the production chain (e.g., integrating primary production with the phases of product processing and marketing), often favoring specific districts with effects that are not always positive for the environment. At the same time, policies to combat poverty and inequality in economically-marginal areas are an essential prerequisite for eliminating territorial imbalances because, only in a broader context of social equity is it possible to stimulate environmental awareness and adoption of sustainable behaviors and practices (Kairis et al., 2013a; Karamesouti et al., 2015; Rontos et al., 2016). In this sense, fostering the permanent dissemination of information also plays an important role in training new generations aware of the potential limitations of their territory (Briassoulis, 2005).

For example, indirect mitigation of land degradation can be left to measures contained in the Common Agricultural Policy (CAP). EU directives have often had negative consequences, particularly for those areas most vulnerable to land degradation (Colantoni et al., 2015b) because they were designed to pursue objectives of general interest and, for this reason, did not take account of specific territorial features. Among the measures foreseen in the new CAP, those aimed at the conservation of rural landscapes should stimulate sustainable development and not only promotion of specific cultural targets, types of farms or groups of stakeholders (Salvati and Carlucci, 2010; Biasi et al., 2017; Duvernoy et al., 2018). Furthermore, support measures should also reconsider the role of local agricultural traditions (typical products, cultivation techniques, land management) as the result of adaptation to land conditions (Kairis et al., 2013a; Salvati and Carlucci, 2014; Zitti et al., 2015).

The participation of local actors in strategic planning and territorial policies is another crucial point in the fight against land degradation (Wilson and Juntti, 2005; Colantoni et al., 2016). This has been found in several cases in Greece and Spain on the involvement of local stakeholders in the definition of planning strategies (Loumou et al., 2000; Iosifides and

Politidis, 2005). However, it has been observed that stakeholders often have a heterogeneous perception of land degradation and, in many cases, they show feelings of 'powerlessness'. It is therefore appropriate to involve local actors from preliminary steps (Patel et al., 2007), involving them in the 'institutional' dialogue together with the political and scientific parties (Kok et al., 2004).

3.7. Mitigation of Land Degradation

The countries of the European Union and the European Community itself, have joined the United Nations Convention to Combat Desertification (UNCCD) along with other industrialized countries such as Canada and Japan, all African and Latin American countries. Italy ratified its participation in the UNCCD in 1997 and subsequently hosted the first Conference of the Parties (decision-making body of the Convention). The objective of the Convention is to 'combat' desertification (understood as a serious and irreversible form of land degradation) and mitigate the effects of drought. This target is particularly appropriate for countries affected by desertification, in particular those in Africa, and it is realized through international cooperation and partnership interventions, within the framework of an approach consistent with Agenda 21: 'Achieving this objective will mean implementing integrated long-term strategies, which simultaneously focus on improving the productivity of land, and on the rehabilitation, conservation and sustainable management of land and water in areas affected by desertification, leading to an improvement in living conditions, at Community level' (UNCCD, Art. 2.2).

The Convention contains four annexes devoted to local specificities and the elaboration of regional Programmes to be developed in Africa, Asia, Latin America, the Caribbean and the North Mediterranean areas. Annex 4 concerns the latter region: Greece, Italy, Portugal and Spain. The document highlights the role played by agriculture, urban development and tourism expansion in triggering desertification processes, with emphasis on rural and marginal areas (Cuadrado-Ciuraneta et al., 2017). The annex also states that the various countries will commit themselves to drawing up action Programmes at national and regional level, establishing the relevant coordination mechanisms and collaborative relationships with other regions (National Committee to Fight Drought and Desertification 1999). Italy, in compliance with its commitments and as an affected country, has designated a specific institutional body to draw up and implement national action Programmes to combat desertification and mitigate the effects of drought (Salvati et al., 2009). The CNLSD, a collegial body of institutional nature, made up of representatives of various ministries, public institutions, research bodies and organizations institutionally involved in activities to combat desertification, was formally established at the Ministry of the Environment by a Prime Minister's Decree dated 26 September 1997. At the same time, the National Action Programme to Combat Drought and Desertification (PAN) was launched.

The CNLSD directly coordinates the implementation of the Convention and one of its objectives is to identify strategies and priorities, within the framework of sustainable development plans and policies, for combating desertification and mitigating droughts. Over the years, the CNLSD has prepared and edited several Programmes aimed essentially at deepening scientific knowledge, spreading an environmental culture and concretely experimenting with various methods to combat land degradation and desertification (Rubio et al., 2009; Kairis et al., 2013b; Kosmas et al., 2013). There are the Operative Plans for the years 2004, 2005 and 2006, supported by the Ministry of the Environment, within which agreements have been stipulated with Universities and Public Research Bodies with the aim of producing Regional and Local Action Plans, as well as a valid scientific and informative documentation. Unfortunately, although the work of the CNLSD has been largely supported by the public research and universities, the methodological approach to desertification assessment has not yet been defined along a clear and shared path, especially about the choice of the set of indicators to be used and the threshold values to be considered.

References

Aldy, J. E. (2006). Per capita carbon dioxide emissions: convergence or divergence? *Environmental and Resource Economics*, 33(4), 533-555.

Arrow, K., Bolin, B., Costanza, R., Dasgupta, P., Folke, C., Holling, C. S., Jansson, B-O., Levin, S., Mäler, K-G., Perrings, C., and Pimentel, D. (1995). Economic Growth, Carrying Capacity, and the Environment. *Science*, 268, 520-521.

Arshad, M. A., and Martin, S. (2002). Identifying critical limits for soil quality indicators in agro-ecosystems. *Agriculture, Ecosystems Environment*, 88(2), 153-160.

Bajocco, S., De Angelis, A., and Salvati, L. (2012). A satellite-based green index as a proxy for vegetation cover quality in a Mediterranean region. *Ecological Indicators*, 23, 578-587.

Barro, R. J., and Sala-i-Martin, X. (2004). *Economic growth*. MIT Press, Cambridge, Massachusetts, USA.

Beato, F. (1998). *Global environmental risk and change. Paths of environmental sociology*. FrancoAngeli, Milan.

Biasi, R., Brunori, E., Ferrara, C., and Salvati, L. (2017). Towards sustainable rural landscapes? a multivariate analysis of the structure of traditional tree cropping systems along a human pressure gradient in a Mediterranean region. *Agroforestry Systems*, 91(6), 1199-1217.

Biasi, R., Colantoni, A., Ferrara, C., Ranalli, F., and Salvati, L. (2015). In-between sprawl and fires: Long-term forest expansion and settlement dynamics at the wildland-urban interface in Rome, Italy. *International Journal of Sustainable Development and World Ecology*, 22(6), 467-475.

Blaikie, P., and Brookfield, H. C. (1987). *Land degradation and society*. Methuen, London.

Borzel, T. A. (2000). Why there is no southern problem: on environmental leaders and laggards in the European Union. *Journal of European Public Policies,* 7(1), 141-162.

Briassoulis, H. (2004). The institutional complexity of environmental policy and planning problems: the example of Mediterranean desertification. *Journal of Environmental Planning and Management,* 47, 115-135.

Briassoulis, H. (2005). *Policy integration for complex environmental problems*. Ashgate, Aldershot.

Carlucci, M., Grigoriadis, E., Rontos, K., and Salvati, L. (2017). Revisiting a Hegemonic Concept: Long-term 'Mediterranean Urbanization' in Between City Re-polarization and Metropolitan Decline. *Applied Spatial Analysis and Policy*, 10, 3, 347-362.

Casadio Tarabusi, E., and Palazzi, P. (2004). An index for sustainable development. *Moneta e Credito LVII*, 226, 123-150.

Ceccarelli, T., Bajocco, S., Perini, L., and Salvati, L. (2014). Urbanisation and land take of high-quality agricultural soils - Exploring long-term land-use changes and land capability in Northern Italy. *International Journal of Environmental Research,* 8(1), 181-192.

Cimini, D., Tomao, A., Mattioli, W., Barbati, A., and Corona, P. (2013). Assessing impact of forest cover change dynamics on high nature value farmland in Mediterranean mountain landscape. *Annals of Silvicultural Research*, 37(1), 29-37.

Colantoni, A., Ferrara, C., Perini, L., and Salvati, L. (2015b). Assessing trends in climate aridity and vulnerability to soil degradation in Italy. *Ecological Indicators,* 48, 599-604.

Colantoni, A., Grigoriadis, E., Sateriano, A., Venanzoni, G., and Salvati, L. (2016). Cities as selective land predators? A lesson on urban growth, deregulated planning and sprawl containment. *Science of the Total Environment*, 545-546, 329-339.

Colantoni, A., Mavrakis, A., Sorgi, T., and Salvati, L. (2015a). Towards a 'polycentric' landscape? Reconnecting fragments into an integrated network of coastal forests in Rome. *Rendiconti Lincei*, 26, 615-624.

Conacher, A. J. (2000). *Land degradation*. Kluwer Academic Publishers, Dordrecht.

Corona, P., Ascoli, D., Barbati, A., Bovio, G., Colangelo, G., Elia, M., and Lovreglio, R. (2014). Integrated forest management to prevent wildfires under mediterranean environments. *Annals of Silvicultural Research*, 38(2), 24-45.

Cuadrado-Ciuraneta, S., Durà-Guimerà, A., and Salvati, L. (2017). Not only tourism: unravelling suburbanization, second-home expansion and 'rural' sprawl in Catalonia, Spain. *Urban Geography*, 38(1), 66-89.

Delfanti, L., Colantoni, A., Recanatesi, F., Bencardino, M., Sateriano, A., Zambon, I., and Salvati, L. (2016). Solar plants, environmental degradation and local socioeconomic contexts: A case study in a Mediterranean country. *Environmental Impact Assessment Review*, 61, 88-93.

Di Feliciantonio, C., and Salvati, L. (2015). 'Southern' Alternatives of Urban Diffusion: Investigating Settlement Characteristics and Socio-Economic Patterns in Three Mediterranean Regions. *Tijdschrift voor Economische en Sociale Geografie*, 106(4), 453-470.

Dinda, S. (2004). Environmental Kuznets curve hypothesis: a survey. *Ecological Economics,* 49, 431- 455.

Duvernoy, I., Zambon, I., Sateriano, A., and Salvati, L. (2018). Pictures from the other side of the fringe: Urban growth and peri-urban agriculture in a post-industrial city (Toulouse, France). *Journal of Rural Studies,* 57, 25-35.

Ezcurra, R. (2007). Distribution dynamics of energy intensities: a cross-country analysis. *Energy Policy*, 35(10), 5254-5259.

Galeotti, M. (2007). Economic growth and the quality of the environment: taking stock. *Environment, Development, Sustainability*, 9, 427-454.

Ibanez, J., Martinez Valderrama, J., and Puigdefabregas, J. (2008). Assessing desertification risk using system stability condition analysis. *Ecological Modelling,* 213, 180-190.

Iosifides, T., and Politidis, T. (2005). Socioeconomic dynamics, local development and desertification in western Lesvos, Greece. *Local Environment,* 10, 487-499.

Jorgens, H. (2005). Diffusion and convergence of environmental policies in Europe. *European Environment,* 15, 61-62.

Kairis, O., Karavitis, C., Kounalaki, A., Salvati, L., and Kosmas, C. (2013a). The effect of land management practices on soil erosion and land desertification in an olive grove. *Soil Use and Management*, 29(4), 597-606.

Kairis, O., Kosmas, C., Karavitis, C., Ritsema, C., Salvati, L., Acikalin, S., Alcalá, M., Alfama, P., Atlhopheng, J., Barrera, J., Belgacem, A., Solé-Benet, A., Brito, J., Chaker, M., Chanda, R., Coelho, C., Darkoh, M., Diamantis, I., Ermolaeva, O., Fassouli, V., Fei, W., Feng, J., Fernandez, F., Ferreira, A., Gokceoglu, C., Gonzalez, D., Gungor, H., Hessel, R., Juying, J., Khatteli, H., Khitrov, N., Kounalaki, A., Laouina, A., Lollino, P., Lopes, M., Magole, L., Medina, L., Mendoza, M., Morais, P., Mulale, K., Ocakoglu, F., Ouessar, M., Ovalle, C., Perez, C., Perkins, J., Pliakas, F., Polemio, M., Pozo, A., Prat, C., Qinke, Y., Ramos, A., Ramos, J., Riquelme, J., Romanenkov, V., Rui, L., Santaloia, F., Sebego, R., Sghaier, M., Silva, N., Sizemskaya, M., Soares, J., Sonmez, H., Taamallah, H., Tezcan, L., Torri, D., Ungaro, F., Valente, S., de Vente, J., Zagal, E., Zeiliguer, A., Zhonging, W., and Ziogas, A. (2013b). Evaluation and Selection of Indicators for Land Degradation and Desertification Monitoring: Types of Degradation, Causes, and Implications for Management. *Environmental Management*, 54(5), 971-982.

Kairis, O., Karavitis, C., Salvati, L., Kounalaki, A., and Kosmas, K. (2015). Exploring the impact of overgrazing on soil erosion and land degradation in a dry Mediterranean agro-forest landscape (Crete, Greece). *Arid Land Research and Management*, 29(3), 360-374.

Karamesouti, M., Detsis, V., Kounalaki, A., Vasiliou, P., Salvati, L., and Kosmas, C. (2015). Land-use and land degradation processes affecting soil resources: Evidence from a traditional Mediterranean cropland (Greece). *Catena*, 132, 45-55.

Kazemzadeh-Zow, A., Zanganeh Shahraki, S., Salvati, L., and Samani, N. N. (2017). A spatial zoning approach to calibrate and validate urban growth models. *International Journal of Geographical Information Science*, 31(4), 763-782.

Kok, K., Rothman, D. S., and Patel, M. (2004). Multi-scale narratives from an IA perspective: Part I. European and Mediterranean scenario development. *Futures*, 38, 261-284.

Kosmas, C., Kairis, O., Karavitis, C., Ritsema, C., Salvati, L., Acikalin, S., Alcalá, M., Alfama, P., Atlhopheng, J., Barrera, J., Belgacem, A., Solé-Benet, A., Brito, J., Chaker, M., Chanda, R., Coelho, C., Darkoh, M., Diamantis, I., Ermolaeva, O., Fassouli, V., Fei, W., Feng, J., Fernandez, F., Ferreira, A., Gokceoglu, C., Gonzalez, D., Gungor, H., Hessel, R., Juying, J., Khatteli, H., Khitrov, N., Kounalaki, A., Laouina, A., Lollino, P., Lopes, M., Magole, L., Medina, L., Mendoza, M., Morais, P., Mulale, K., Ocakoglu, F., Ouessar, M., Ovalle, C., Perez, C., Perkins, J., Pliakas, F., Polemio, M., Pozo, A., Prat, C., Qinke, Y., Ramos, A., Ramos, J., Riquelme, J., Romanenkov, V., Rui, L., Santaloia, F., Sebego, R., Sghaier, M., Silva, N., Sizemskaya, M., Soares, J., Sonmez, H., Taamallah, H., Tezcan, L., Torri, D., Ungaro, F., Valente, S., de Vente, J., Zagal, E., Zeiliguer, A., Zhonging, W., and Ziogas, A. (2013). Evaluation and Selection of Indicators for Land Degradation and Desertification Monitoring: Methodological Approach. *Environmental Management*, 54(5), 951-970.

Loumou, A., Giourga, C., Dimitrakopoulos, P., and Koukoulas, S. (2000). Tourism contribution to agro- ecosystems conservation; the case of Lesbos island, Greece. *Environmental Management*, 26, 363- 370.

Mainguet, M. (1994). *Desertification: natural background and human mismanagement.* Springer, Berlin.

Marchetti, M., Vizzarri, M., Lasserre, B., Sallustio, L., and Tavone, A. (2015). Natural capital and bioeconomy: challenges and opportunities for forestry. *Annals of Silvicultural Research*, 38(2), 62-73.

Mavrakis, A., Papavasileiou, C., and Salvati, L. (2015). Towards (Un) sustainable urban growth? Industrial development, land-use, soil depletion and climate aridity in a Greek agro-forest area. *Journal of Arid Environments*, 121, 1-6.

Munafò, M., Salvati, L., and Zitti, M. (2013). Estimating soil sealing rate at national level - Italy as a case study. *Ecological Indicators*, 26, 137-140.

Neumayer, E. (2001). The human development index and sustainability—a constructive proposal. *Ecological Economics*, 39(1), 101-114.

Patel, M., Kok, K., and Rothman, D. S. (2007). Participatory scenario construction in land-use analysis: an insight into the experiences created by stakeholder involvement in the Northern Mediterranean. *Land Use Policy,* 24, 546-561.

Pili, S., Grigoriadis, E., Carlucci, M., Clemente, M., and Salvati, L. (2017). Towards sustainable growth? A multi-criteria assessment of (changing) urban forms. *Ecological Indicators*, 76, 71-80.

Pisani-Ferry, J. (2005). *An agenda for a growing Europe.* Oxford University Press, Oxford, UK.

Rontos, K., Grigoriadis, E., Sateriano, A., Syrmali, M., Vavouras, I., and Salvati, L. (2016). Lost in protest, found in segregation: Divided cities in the light of the 2015 'Oχι' referendum in Greece. *City, Culture and Society,* 7(3), 139-148.

Rubio, J. L., Safriel, U., Blum, W. E. H., and Pedrazzini, F. (2009). *Water scarcity, land degradation and desertification in the Mediterranean region*. Springer, Heidelberg.

Salvati, L. (2013). Monitoring high-quality soil consumption driven by urban pressure in a growing city (Rome, Italy). *Cities*, 31, 349-356.

Salvati, L. (2014). Agro-forest landscape and the 'fringe' city: A multivariate assessment of land-use changes in a sprawling region and implications for planning. *Science of the Total Environment*, 490, 715-723.

Salvati, L., and Carlucci, M. (2010). Estimating land degradation risk for agriculture in Italy using an indirect approach. *Ecological Economics*, 69(3), 511-518.

Salvati, L., and Carlucci, M. (2011). The economic and environmental performances of rural districts in Italy: Are competitiveness and sustainability compatible targets? *Ecological Economics*, 70(12), 2446-2453.

Salvati, L., and Carlucci, M. (2014). A composite index of sustainable development at the local scale: Italy as a case study. *Ecological Indicators*, 43, 162-171.

Salvati, L., and Zitti, M. (2005). Land degradation in the Mediterranean Basin: Linking bio-physical and economic factors into an ecological perspective. *Biota,* 6(43132), 67-77.

Salvati, L., and Zitti, M. (2009). Substitutability and weighting of ecological and economic indicators: Exploring the importance of various components of a synthetic index. *Ecological Economics*, 68(4), 1093-1099.

Salvati, L., and Zitti, M. (2012). Monitoring vegetation and land-use quality along the rural-urban gradient in a Mediterranean region. *Applied Geography*, 32(2), 896-903.

Salvati, L., Petitta, M., Ceccarelli, T., Perini, L., Di Battista, F., and Scarascia, M. E. V. (2008a). Italy's renewable water resources as estimated on the basis of the monthly water balance. *Irrigation and Drainage*, 57(5), 507-515.

Salvati, L., Zitti, M., and Ceccarelli, T. (2008b). Integrating economic and environmental indicators in the assessment of desertification risk: A case study. *Applied Ecology and Environmental Research*, 6(1), 129-138.

Salvati, L., Zitti, M., Ceccarelli, T., and Perini, L. (2009). Developing a synthetic index of land vulnerability to drought and desertification. *Geographical Research,* 47(3), 280-291.

Salvati, L., Bajocco, S., Ceccarelli, T., Zitti, M., and Perini, L. (2011). Towards a process-based evaluation of land vulnerability to soil degradation in Italy. *Ecological Indicators*, 11(5), 1216-1227.

Salvati, L., Gemmiti, R., and Perini, L. (2012a). Land degradation in Mediterranean urban areas: An unexplored link with planning? *Area,* 44(3), 317-325.

Salvati, L., Perini, L., Sabbi, A., and Bajocco, S. (2012b). Climate Aridity and Land-use Changes: A Regional-Scale Analysis. *Geographical Research*, 50(2), 193-203.

Salvati, L., Tombolini, I., Perini, L., and Ferrara, A. (2013a). Landscape changes and environmental quality: The evolution of land vulnerability and potential resilience to degradation in Italy. *Regional Environmental Change*, 13(6), 1223-1233.

Salvati, L., Morelli, V. G., Rontos, K., and Sabbi, A. (2013b). Latent exurban development: City expansion along the rural-to-urban gradient in growing and declining regions of southern Europe. *Urban Geography*, 34(3), 376-394.

Salvati, L., Sateriano, A., and Grigoriadis, E. (2016). Crisis and the city: profiling urban growth under economic expansion and stagnation. *Letters in Spatial and Resource Sciences*, 9(3), 329-342.

Smiraglia D., Ceccarelli T., Bajocco S., Salvati L., and Perini L. (2016). Linking trajectories of land change, land degradation processes and ecosystem services. *Environmental Research*, 147, 590-600.

Steer, A. (1998). Making development sustainable. *Advances in Geo-Ecology*, 31, 857-865.

Stern, D. I. (2004). The rise and fall of the Environmental Kuznets Curve. *World Development*, 8, 1419- 1439.

Thornes, J. B. (2004). Stability and instability in the management of Mediterranean desertification. In: J. Wainwright and M. Mulligan (eds.) *Environmental modelling: finding simplicity in complexity*. Wiley, Chichester, UK.

Tumpel-Gugerell, G., and Mooslechner, P. (2003*). Economic convergence and divergence in Europe. Growth and regional development in an enlarged European union*. Edward Elgar, Chichester.

Wilson, G. A., and Juntti, M. (2005). *Unravelling desertification: policies and actor networks in Southern Europe.* Wageningen, Wageningen Academic Publishers.

Zambon, I., Benedetti, A., Ferrara, C., and Salvati, L. (2018). Soil Matters? A Multivariate Analysis of Socioeconomic Constraints to Urban Expansion in Mediterranean Europe. *Ecological Economics*, 146, 173-183.

Zambon, I., Serra, P., Sauri, D., Carlucci, M., and Salvati, L. (2017). Beyond the 'mediterranean city': Socioeconomic disparities and urban sprawl in three Southern European cities. *Geografiska Annaler, Series B: Human Geography*, 99(3), 319-337.

Zitti, M., Ferrara, C., Perini, L., Carlucci, M., and Salvati, L. (2015). Long-term urban growth and land-use efficiency in Southern Europe:

Implications for sustainable land management. *Sustainability* (Switzerland), 7(3), 3359-3385.

Zuindeau, B. (2007). Territorial equity and sustainable development. *Environmental Values*, 16, 253- 268.

In: Land Degradation: The Main Challenge ISBN: 978-1-53615-575-4
Editors: Ilaria Zambon et al.

Chapter 4

MEASURING LAND DEGRADATION: A SOCIODEMOGRAPHIC OUTLOOK

Rita Biasi[1,*], Luca Salvati[2,†], Pere Serra[3,‡] and Ilaria Zambon[1]

[1]Tuscia University, Viterbo, Italy
[2]Council for Agricultural Research and Economics (CREA), Arezzo, Italy
[3]Universitat Autònoma de Barcelona, Barcelona, Spain

Measurement and classification of a given phenomenon are important, as well as the correct formulation of the initial working hypothesis. One of the main problems in the measurement theory is the conceptualization into simpler components suitable to formulate interpretations for subsequent analysis (operationalization). The choice of variables for analyzing a given phenomenon represents a crucial condition for a coherent understanding of complex issues and will be even more valid when it provides information with respect to the initial hypothesis. To define variables, it is necessary to

* Corresponding Author's E-mail: biasi@unitus.it; ilaria.zambon@unitus.it
† Author's E-mail: luca.salvati@crea.gov.it
‡ Author's E-mail: pere.serra@uab.cat

have adequate information in terms of coherence, reliability and completeness, to allow the construction of relevant indicators useful to represent a given phenomenon.

4.1. Logical Schemes

One of the most widely used methods of environmental analysis and reporting is the so-called "Determinants, Pressures, State, Impacts and Responses" (DPSIR) scheme, developed by the United Nations Organization for Economic Cooperation and Development (OECD). This scheme, originally conceived as a "Pressure-State Response (PSR)" model, has been widely used in environmental studies developed in international and national contexts to highlight the causal sequence between human pressures, environmental impacts and responses to mitigate these impacts. The DPSIR framework was a European Environment Agency's reference framework for reporting environmental issues, including desertification (Figure 1).

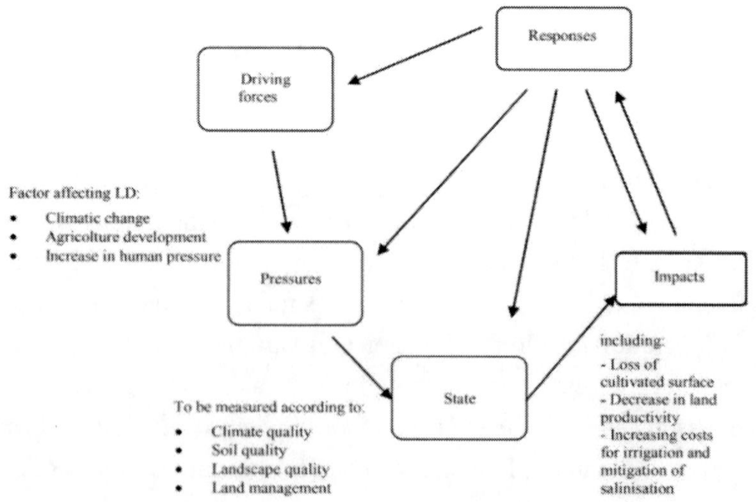

Source: Own elaboration.

Figure 1. A simplified DPSIR scheme applied to land degradation in southern Europe.

The Driving Forces component includes the various factors that trigger environmental problems and generate pressures. These causes are mainly related to human action, including population growth and economic activities such as agriculture, tourism and industry (Cimini et al., 2013; Corona et al., 2014; Marchetti et al., 2015; Cuadrado-Ciuraneta et al., 2017). In recent years, there have been further causes of climate change (Salvati et al., 2012; Colantoni et al., 2015a; Ferrara et al., 2017). The other components, defined as "Pressures," "State" and "Impacts" indicate the effects produced on the environmental system through all known interactions. The "Pressure" component defines the direct (and measurable) effects of Driving Forces (e.g., polluting emissions, urbanization, deforestation, consumption of material and energy resources, loss of natural habitats), while the "State" component is represented by variables that measure the overall quality of the territory (air quality, water quality, level of biodiversity). The "Impacts" are the final effects coming from the Driving Forces that make explicit the cause-effect relations (e.g., reduction of cultivated areas, loss of biodiversity). The "Response" component, finally, refers to human interventions to (directly or indirectly) contrast land degradation such as, for example, the measures included in the Common Agricultural Policy (CAP), the UNCCD research and other environmental interventions. In principle, the DPSIR scheme represents an interpretative reference more than a methodology and aims at highlighting the cause-effect connections by dynamically assessing the impact of Driving Forces according to resilience and adaptation measures.

Brouwer et al. (1991), Puigdefabregas and Mendizabal (1998) and Camci Cetin et al., (2007), indicate complex processes as the principal causes of land vulnerability to desertification in southern Europe (Salvati et al., 2009, 2013b). The increasing human pressure in coastal and lowland areas through urban growth, internal migration, tourism, and agricultural intensification (Colantoni et al., 2015b, 2016; Zitti et al., 2015; Salvati et al., 2016; Cuadrado-Ciuraneta et al., 2017; Kazemzadeh-Zow et al., 2017; Duvernoy et al., 2018), in turn related with unsuitable exploitation of water and soil resources, leads to soil salinization, compacting, sealing and pollution (Loumou et al., 2000; Iosifides and Politidis, 2005a; Atis, 2006;

Salvati et al., 2008a; Munafò et al., 2013; Karamesouti et al., 2015). In marginal, inland areas depopulation and land abandonment together with intense rainfall cause soil deterioration (Garcia Latorre et al., 2001; Hein 2007; Salvati and Zitti 2008a), particularly in sloping areas exposed to water erosion (Salvati et al., 2008a). Owing to the heterogeneity of cause-effect relationships, few studies attempted to assess the latent impact of anthropogenic drivers on landscape quality and land degradation (Mairota et al., 1998). Identifying efficient methodologies which evaluate effectively the state of desertification process is problematic. The statistical and geographically related data are not adequate at all to assure an efficient monitoring of desertification risk even in developed countries (Salvati et al., 2008b). There is scope for additional studies and methodological approaches (Seely and Wohl, 2004).

4.2. Towards Sustainability Indicators

The DPSIR scheme has the advantage of being easily "operationalized" through the identification and/or creation of suitable indicators which, in our case, are both of a socioeconomic and environmental nature. An indicator can be defined as a summary measure, usually expressed in a quantitative form, coinciding with an individual variable over time and space. Depending on the case, this can be a classification, sorting, counting or measuring procedure. The choice of the indicator depends on its capacity to explain a given phenomenon. The logical connections between the phenomenon and its indicator do not always appear perfectly clear or universally shared. In the choice of an indicator there are different criteria that can orient choices even apparently distant from the phenomenon under analysis. It should be noted that, in line with the definitions given, there are no "perfect" indicator but only indicators that meets our criteria or the desired characteristics. This depends on the need to give completeness to the investigation, or on the effective availability of information and, even, on the mere "subjective sensitivity" of research (Figure 2).

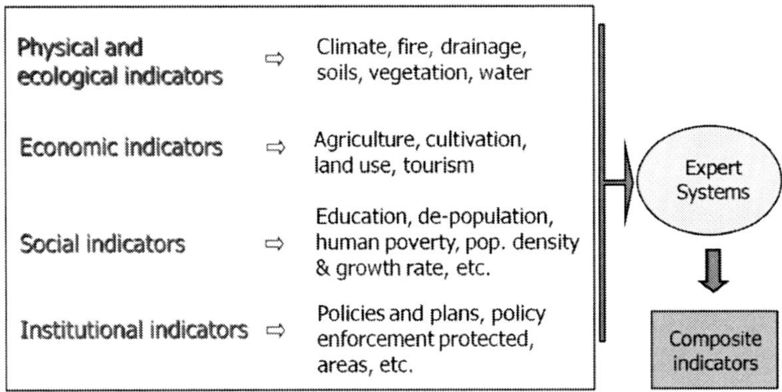

Source: Our elaboration.

Figure 2. Conceptual dimensions of land degradation.

In the specific case of environmental analysis, the notion of an indicator is often associated with that of an "index," which should not be confused, however, since the latter represents a composition of suitable aggregated indicators (OECD, 1994). The OECD and the EEA highlight some properties of the indicators used in the environmental field, with the aim to assess sustainability of socioeconomic activities and the different forms of spatial governance: (i) representativeness of the information bases and adherence to shared standards; (ii) measurability and accuracy; and (iii) policy relevance. In the light of the above considerations, indicators are a suitable tool for assessing medium-term progress in supporting policy adjustment and priority setting (Ceccarelli et al., 2014; Biasi et al., 2015; Zitti et al., 2015; Carlucci et al., 2017).

4.3. Quantitative Assessment of Land Degradation

In the case of land degradation, the multiplicity and diversity of causes are required as a relevant set of indicators (Kairis et al., 2013a, 2013b), many of which, not available from primary information sources, can be effectively derived from secondary information sources (e.g., official statistics). Even considering earlier research on land sensitivity to degradation, a shared approach to the different analysis scales does not yet seem to be fully identified. This can be easily highlighted by different surveys carried out at regional level, where the identification of predisposing factors, the choice of indicators and the procedures for calculating the final index are object of intense debate. This is mainly due to the lack of relevant quantitative and qualitative references, as well as to the inconsistent definition of key notions (such as environmental sustainability) and, in part, to the lack of data (Brandt et al., 2003). It is therefore useful to schematize what has been said so far, classifying conceptual references by means of summary indicators.

Several criteria should be satisfied for selection of relevant indicators including empirical formulation, digital accessibility, statistical reliability and effectiveness, adaptability to multi-temporal analysis and to different spatial scales. This approach, although it could be constrained by the reduced availability of basic data, allows a diachronic overview of a given phenomenon. The result depends on the set of indicators chosen to estimate vulnerability to land degradation as a function of climate change and anthropogenic pressure (Salvati et al., 2012; Colantoni et al., 2015a; Karamesouti et al., 2015). In terms of the appropriate cartography and statistical description of the adopted variables, the information collected will be even more useful when being able to interpret the territorial contexts at the different operational scales. Identifying the relevant scale to effectively bring out the problems of a given territory, or rather the elementary unit (spatial and temporal) to which aggregate the basic information is a relevant analysis' step.

A spatial aggregate is a geographical unit which generally coincides with administrative boundaries (e.g., region, province, municipality) or physiographic-administrative boundaries (e.g., river basin, agricultural region, mountain community). The spatial unit depends on the coherency of the geographical scale with the purpose of the study. Spatial scale is an element of analysis that concerns numerous fields of research, from demography to economics, from environmental sciences to geophysics. The possibility of making diachronic comparisons has recently been favored by recovery and validation of numerous historical time series from census sources and annual or periodic surveys. For instance, the choice of the municipal scale allows use of indicators from official census sources (General Censuses of Population, Buildings, Industry, Agriculture). The time scale is linked to the frequency of the census surveys. A problematic dimension of the empirical analysis is represented by data heterogeneity resulting from the use of different information sources, which can bring to a high degree of redundancy. In this regard, various methodologies have been proposed for the construction of composite indicators that make use of multivariate statistical techniques removing (or containing) redundancy (Salvati, 2014; Biasi et al., 2017; Zambon et al., 2018). These approaches, initially used for the dimensional reduction of complex data matrices, allow implicit assessment of the importance of individual variables in quantitative datasets. This approach also provides unified analysis of socioeconomic and biophysical indicators.

4.3.1. Monitoring Environmental Conditions

The use of the DPSIR model described above, is an advanced approach to the study of land degradation and desertification. Based on the United Nations Convention to Combat Desertification (UNCCD), a series of initiatives were planned at supranational, national and local levels. The UNCCD, through its "Committee on Science and Technology," worked on the definition of benchmarks and indicators to encourage the adoption of widely shared guidelines. For instance, in Italy, the Convention was implemented by the National Committee for the Fight against Desertification (CNLD), which in 1999 established a national reference framework summarized in the "National Communication to Combat Drought and Desertification" (National Committee to Combat Drought and Desertification 1999). The Committee carried out a comprehensive analysis at national level to identify areas sensitive to desertification, producing a thematic map at a 1: 250,000 scale. For the identification of sensitive areas, climatic (aridity index), soil (soil and climate index), vegetation (Corine Land Cover cartography) and anthropogenic (demography) factors were considered (e.g., Garcia Latorre et al., 2001;

Bajocco et al., 2012; Salvati et al., 2013c; Ceccarelli et al., 2014; Salvati and Ferrara, 2014; Colantoni et al., 2015a; Smiraglia et al., 2015). Based on such analysis, the national territory has been classified into different areas that are prone to desertification (Figure 3).

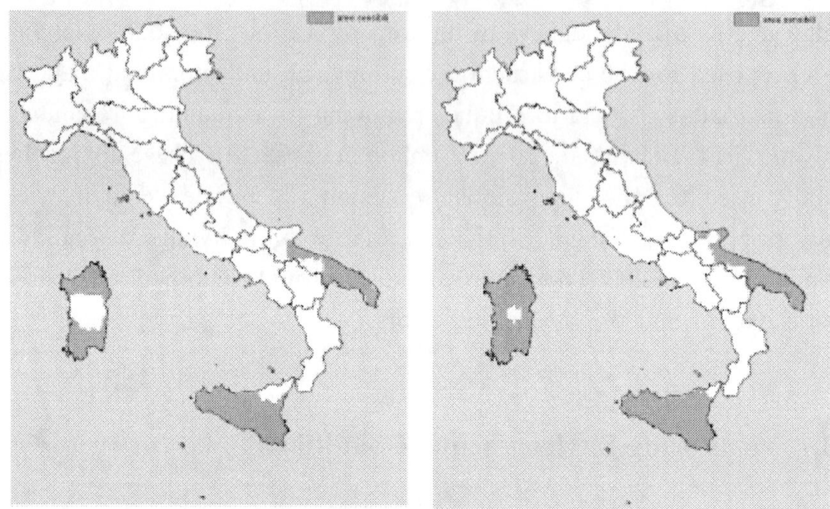

Source: Our elaboration.

Figure 3. Areas prone to desertification in 1990 (left) and 2000 (right) in Italy.

Although approaches based on multidimensional analysis of land degradation are recognized as a reliable methodology (Feoli et al., 2003; Kairis et al., 2015; Karamesouti et al., 2015), scientific literature highlights the lack of diachronic analysis, especially regarding socioeconomic aspects (Rontos et al., 2016; Delfanti et al., 2016; Zambon et al., 2017, 2018). This is mainly due to the lack of diachronic data sets (Basso et al., 2000; D'Angelo et al., 2000; Salvati and Zitti, 2009; Salvati et al., 2012). Among the most widespread computational procedures usually adopted at the international scale, it is worth mentioning the ESA (Environmentally Sensitive Areas) scheme, which appears to be the most suitable for the study of the Mediterranean basin and other semi-arid environments of North Africa and the Middle East (Salvati et al., 2013a). Recent studies (e.g., Fraser et al., 2006) have underlined the effectiveness of the ESA system developed in the framework of the MEDALUS, DESERTLINKS,

MEDACTION and DESIRE projects. This is due to its simplicity in the construction of the model and its flexibility in the use of relevant indicators. The final index resulting from the application of the ESA scheme can assess the level of land degradation of a given territory by analyzing four thematic dimensions into which biophysical and socioeconomic indicators have been classified (Kosmas et al., 2003). The methodology introduces a two-step evaluation process. In the first phase, the elementary data (for the individual variables used) are combined to provide four quality indicators: A Climate Quality Index, a Soil Quality Index, a Vegetation Quality Index and a land Management Quality Index. In the second phase, the Environmentally Sensitive Area Index (ESAI) is calculated, for each spatial unit considered, as the geometric mean of the four partial indicators of environmental quality.

The ESA index assumes values that vary from 1 to 2: the lowest scores indicate negligible or low vulnerability to land degradation, while the highest scores indicate critical environmental conditions that can determine a high level of vulnerability. According to Basso et al., (2000), different areas can be classified as "not affected" if the ESAI is below 1.175; "potentially affected (1.175 - 1.225); "fragile" (1.225 - 1.375); "critical," if ESAI exceeds 1.375. This classification was determined by empirical observation of the index statistical distribution in different experimental sites and by cross analysis with supplementary indicators. Such classifications represent an indirect and relative estimate that is influenced by the weight attributed to the different qualities considered: the selection of the informative layers and the calculation of the thematic indicators are an open process, as the choice of the associated variables and weights can vary according to the operational context.

While assessing the conditions of environmental degradation, the ESAI can be used as an early-warning tool since it identifies conditions that have the potential to evolve into irreversible states of desertification. In other words, the ESA scheme does not imply a process based on the assessment of individual specific degradation processes (e.g., soil erosion, salinization, compaction, sealing) but rather evaluates the different contextual factors that trigger land degradation (Karamesouti et al., 2015).

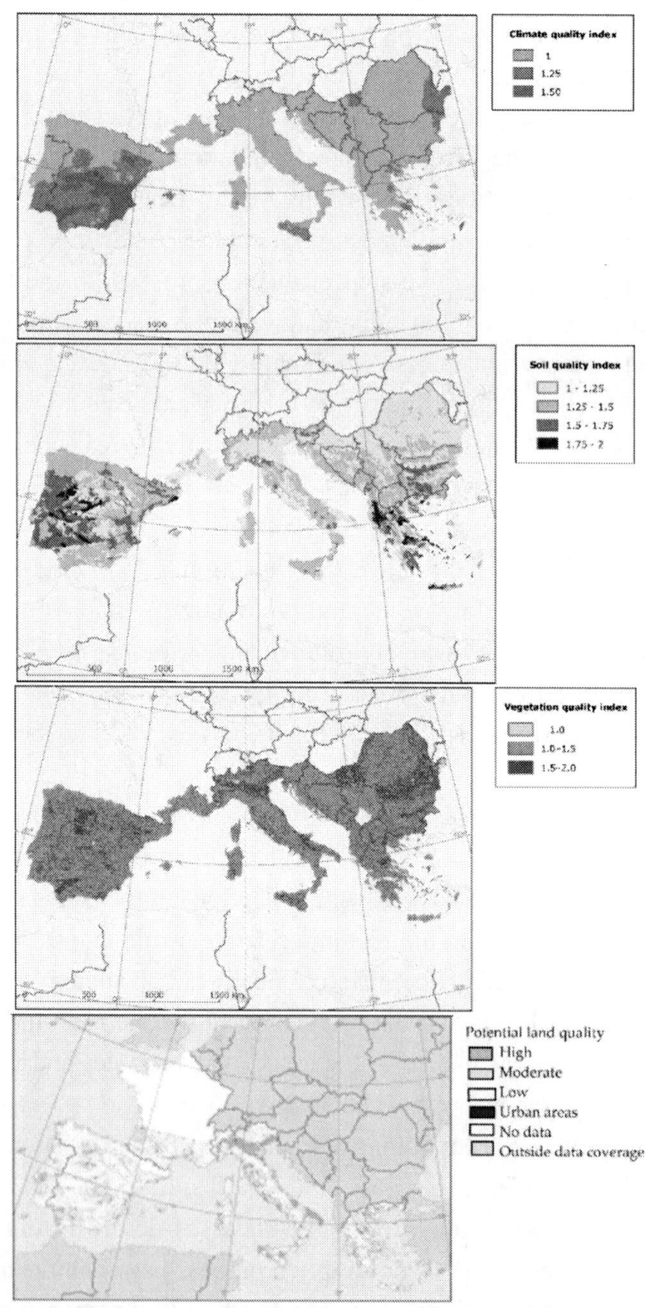

Source: EEA.

Figure 4. Application of the ESAI to monitor desertification risk in Europe.

Table 1. Selected procedures for desertification monitoring

Procedure	Indices	Aggregation Method
Environmental Sensitive Areas	ESA based on three dimensions	Geometric average after variable normalization
	Standard ESA based on four dimensions	Geometric average after variable normalization
	ESA index expressed in relative terms	Geometric mean after variable transformation (2 steps)
	A review of ESA applications	Various procedures
	Modified ESA index for anthropogenic pressures	Geometric mean after standard-setting of variables and their weighing
	Modified ESA index for agricultural mechanization	Geometric mean after standard-setting of variables and their weighing
	Modified ESA index	Geometric mean after standard-setting of variables and their weighing
Remote sensing	Desertification Risk Index (DRI)	Arithmetic mean
Neural Network	Drought Sensitivity Index (DSI)	Analysis of historical series
Geostatistic	Severity indicator	Geostatistical analysis
GIS	Various server indicators	Environmental indicators
DPSIR	Composite Index	Environmental indicators
System stability condition	Composite Index	Output analysis
Multi-way Analysis	Land Vulnerability Index (LVI)	Arithmetic mean after normalization of the variables and their weighing
Statistical multivariate Analysis	Risky regions	Multivariate strategy with three different techniques

Source: Own elaboration.

This approach is particularly useful for diachronic comparisons. Finally, the ESA methodology represents the standard currently in use at the European Environment Agency for environmental reporting on a continental scale (Figure 4). When measurements are needed to evaluate the level of land vulnerability (Salvati et al., 2009, 2013b), the Environmentally Sensitive Area (ESA) framework appears to be the most applied in the Mediterranean basin (Fraser et al., 2006; Sepehr et al., 2007; Ali and Baroudy 2008; Spilanis et al., 2008; Lavado Contador et al., 2009; Bajocco et al., 2012; Salvati et al., 2013a; Zitti et al., 2015).

Further approaches to estimate land degradation have recently been introduced as alternative or complementary methods to the ESAI (Hill et al., 2008). Some of these procedures are presented in Table 1. The procedures in Feoli et al., (2003) have proven to be useful for monitoring the ecological conditions which lead to drought and desertification in arid lands with limited plant cover. A simplified index was proposed, allowing a classification of land sensitivity on a municipal basis, exploiting census data from official statistics, focusing on socioeconomic variables (Iosifides and Politidis, 2005; Salvati et al., 2008; Salvati and Zitti 2009; Salvati and Carlucci, 2011; Di Feliciantonio and Salvati, 2015; Delfanti et al., 2016; Rontos et al., 2016). More recently, some studies introduced a Land Vulnerabiliy Index (LVI), obtained from a large set of environmental indicators collected at municipal level and processed using multi-way analysis. The advantages of LVI are that (i) it is flexible enough to consider every variable that is involved in land degradation processes (Karamesouti et al., 2015) and (ii) the relative procedure gives an objective weight to each input, estimating the importance of each factor.

The LVI thus tends to exceed the major limit of the ESA standard model, which is precisely the subjective attribution of variable weights, implying the risk of underestimating the role of specific factors (e.g., climate variability, precipitation concentration or the available water capacity of the soil) which can be decisive to assess land vulnerability (Salvati et al., 2008a; Savo et al., 2012). The LVI was first applied at country scale, where nearly 70 indicators (including 17 indicators for climate change, 5 for soil sealing and urbanization, 5 for soil salinization,

22 for soil erosion, 4 for soil pollution, and 18 for agriculture) were selected to represent 6 different degradation systems. Within each degradation system, the ratio between each indicator and its impact on land quality was defined through multivariate statistical procedures and contributes to define composite risk index. Statistical and modelling techniques are considered the most useful way to interpret the dynamic interactions between economic, social and natural capitals and degradation processes. However, the use of formal models does not ensure more accurate and comprehensive results than those obtained from qualitative studies based on interviews with privileged witnesses or grounded on a visual landscape analysis. On the contrary, such approaches give strength to formal models and offer interpretative hints enrichening quantitative analyses (Makzhoumi, 1997; Loumou et al., 2000; Iosifides and Politidis, 2005). Models also suffer limitations due to the availability of homogeneous and comparable statistical data in time and space, which makes it difficult to evaluate the role played by different socioeconomic contexts in desertification issues (Delfanti et al., 2016; Zambon et al., 2017, 2018).

4.3.2. Monitoring Socioeconomic Aspects

Quantitative approaches, specifically addressing socioeconomic issues, can be classified according to methodology (e.g., exploratory, interpretative, simulation, modelling), economic scale (e.g., micro-economic or macro-economic), geographical scale (local, regional, country, continental or global) and time scale (Iosifides and Politidis, 2005; Ceccarelli et al., 2014; Biasi et al., 2015; Di Feliciantonio and Salvati, 2015; Zitti et al., 2015; Delfanti et al., 2016; Carlucci et al., 2017). As far as exploratory or interpretative statistical analysis is concerned, the most widely used techniques include regression analysis, with extensions taking account of spatial and temporal dimensions, non-parametric analysis and multivariate techniques exploring large data matrices (Salvati, 2014; Biasi et al., 2017; Zambon et al., 2018).

Microeconomic models explain how individuals allocate their resources (including natural capital), and use endogenous socioeconomic variables describing preferences, price levels, institutions, access to infrastructure and services, as well as technological alternatives. An important distinction should be made between models that assume prices as they are determined by the market (perfect competition) and those that assume fixed or variable prices but not determined directly by the market. Within the first category, production decisions are driven by market prices and can be studied as a problem of profit maximization. In the latter case, prices are instead determined subjectively and endogenously through shadow values in markets that are not perfectly competitive. Factors such as the allocation of resources and the household composition are important, as the consumption aspect must be considered at the time of the production decision. This distinction is fundamental when considering spatial aspects as land-use, consumption levels, and natural resource degradation mechanisms. These may vary according to population growth, changes in agricultural prices, income levels, and many other variables (Salvati, 2013, 2014).

Regional models represent interpretative schemes addressing the multiple aspects that characterize ecosystem dynamics, economic growth and the social context. These approaches also consider institutional, cultural, political and socioeconomic variables together with environmental factors (Iosifides and Politidis, 2005; Huby et al., 2007; Di Feliciantonio and Salvati, 2015; Delfanti et al., 2016; Rontos et al., 2016; Zambon et al., 2017, 2018). This seems to be relevant for the study of land degradation as the most important consequences are observed at the district (or regional) level, while the decisions taken at the level of individual territorial actors (studied by the micro-economic models discussed above), appear to be much more difficult to integrate into the reference environmental context. Such models generally have a spatial coverage limited to a region with specific characteristics in landscape, agriculture, political history, and land-use patterns.

Most regional models use spatial or non-spatial regression tools (Nijkamp, 1999). For example, some models indirectly incorporate a spatial approach by measuring land-use impacts of variables such as topography, soil quality, rainfall, population density. This analysis has improved using Geographic Information Systems (GIS) that have made it easier to manipulate geographic data. In these models, however, changes in the behavior of economic agents may be difficult to model (Perez-Trejo and Clarke, 1996). However, the non-spatial approach remains even more widely used, starting from disaggregated data with regional coverage and comparative purposes (Anselin, 1995). In both cases, the quality of the data appears to be a relevant factor; in this direction, there is a growth in applications that use alongside official statistical data and remote sensing information, often in combination with field surveys.

Macroeconomic models, generally developed at national or supranational level, underline the relationship between basic variables, decision parameters, and the environment. From an analytical point of view, the integration between stochastic simulation and regression models is suitable for these models. However, to represent complex macro-economic processes in a quantitative and formal context, it is necessary to set constraints on the number of variables and to make some a-priori assumptions. These models add important aspects of knowledge with respect to the two types described above, assessing how the decision-making parameters influence the behavior of local actors and, at the same time, how the context variables determine a set of decision-making parameters, providing a link between macroeconomic variables and policy tools. Secondly, most of these models consider in detail the interaction between different production sectors, e.g., the relationship between different subsectors of agriculture and other sectors of the economy, which could be particularly relevant in a structural analysis of the underlying factors of land degradation (Salvati and Zitti, 2008b).

4.4. Estimating The Economic Costs of Land Degradation

The factors identified as influencing land degradation, mutual interactions among critical factors and the magnitude of their effects can vary significantly at both regional and local levels (Antle and Heiderbrinck, 1995; Arrow et al., 1995; Steer, 1998). Models based on individual case studies may provide conflicting results when applied to other areas with different environmental and socioeconomic contexts (Iosifides and Politidis, 2005; Di Feliciantonio and Salvati, 2015; Delfanti et al., 2016; Rontos et al., 2016). The integration with qualitative analysis appears useful to highlight local specificities and relate them, as much as possible, to a broader interpretative scale (Iosifides and Politidis, 2005). A quantitative assessment of the economic costs of environmental degradation is therefore of great interest (Atis, 2006; Hein, 2007; Delfanti et al., 2016). The unsustainable use of natural resources has negative effects on economic systems, causing a decrease in the stock of natural capital and reducing ecosystem services. Although the debate on these issues has intensified in recent years, the monetary aspects of desertification have been studied rarely. The fight against desertification should be carried out through targeted sectoral policies considering the general framework of social context and the surrounding environmental conditions (Briassoulis, 2011). Most of the existing studies at international level focus on the costs of desertification in relation to their impact on agriculture (Figure 5). The following questions are proposed: What are the immediate (short-term) and future (long-term) effects of the costs of land degradation on a country or region? And, more specifically, what are the (future) costs of inaction in an environmental context where climate change is combined with increased human pressure and unsustainable management of rural land?

Unfortunately, there is no direct and immediate answer to these questions (Bojo, 1996). However, the scientific literature on desertification has many points in common with the management of natural resources and

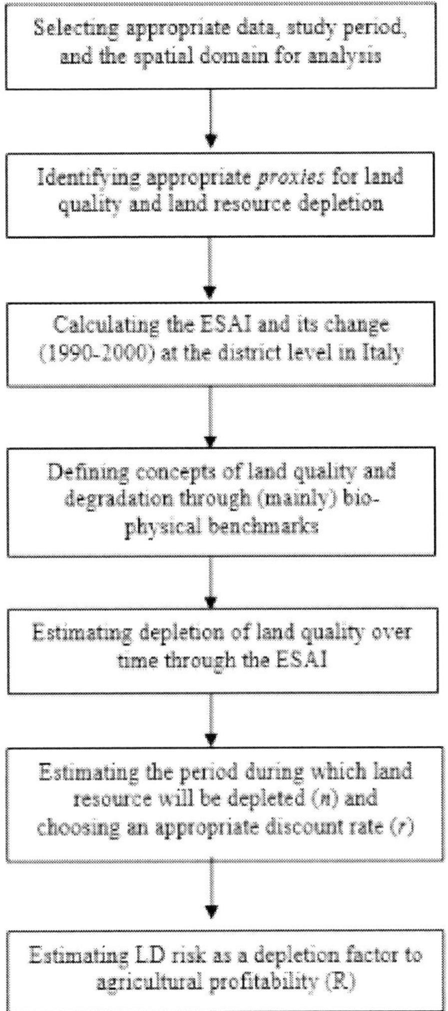

Source: Own elaboration.

Figure 5. A procedure estimating land degradation costs in agriculture.

climate change (Colantoni et al., 2015a; Salvati et al., 2008a, 2012). Therefore, in the economic evaluation of land degradation and the consequent policies of mitigation and adaptation, it is possible to refer to concepts and practices already tested in the study of similar themes (Salvati and Carlucci, 2010). The assessment of the economic costs of land degradation should also consider the so-called costs of inaction, due both to direct impacts in terms of reduction of ecological services offered by degraded land and to indirect impacts (e.g., unemployment and internal migration, but also loss of biodiversity). The assessment of the consequences of inaction must consider the overall costs and benefits of implementing mitigation and/or adaptation policies (APAT, 2006). Some critical issues in this context include identification of reference scenarios for assessment of inaction, management of spatial and temporal cost trends, classification of impacts as (reversible or irreversible) and the importance of different forms and processes of degradation.

There are many possible options for the economic assessment of land degradation that could be grouped into three categories: Cost-Benefit Analysis (CBA), Cost-Effectiveness Analysis (CEA) and Multi-Criteria Analysis (MCA). The first technique involves a monetary assessment of the costs and benefits of interventions, the second compares the costs of various mitigation/adaptation strategies to identify the least costly alternative, and the third is based on identifying the best solution by reference to a pre-defined list of criteria and/or targets that are not necessarily monetary (Bojo, 1996). Once the multiplicity of possible cost definition has been defined, three economic reference concepts can be identified: (i) immediate gross annual loss, (ii) discounted gross future loss and (iii) cumulative discounted gross loss. In the first concept, economic losses are because of land degradation in the recent past. The second concept draws attention to the fact that the loss of land quality in one year will have an impact on the level of production in the following year. The third concept identifies the depletion of land capital as a process of cumulative degradation, observed over a long period.

Different approaches have also been proposed to derive the monetary values linked to degradation of land resources. Basically, the problem lies in the fact that resources are measured in physical units. Bojo (1996) identifies three procedures, respectively called: replacement cost, loss of productivity and protection costs. Based on this theoretical framework, only few studies have attempted to quantify the costs of the main forms of land degradation in Southern Europe. Some of them (e.g., De Groot et al., 2002; Atis, 2006; Hein, 2007) have tried to establish a general framework for the analysis of impacts and to provide a preliminary assessment of socioeconomic costs in specific areas of the Mediterranean basin, by using the criterion of environmental functions (Iosifides and Politidis, 2005; Salvati et al., 2013a; Di Feliciantonio and Salvati, 2015; Delfanti et al., 2016). An environmental function is defined as "the ability of the environment to provide goods and services that satisfy human needs, directly or indirectly" (De Groot et al., 2002). The approach consists of the following steps:

- identification of ecosystems affected by land degradation;
- determination of the key environmental functions of each ecosystem;
- evaluation of the main key functions;
- identification of active degradation processes for each landscape;
- assessment of impact of land degradation on environmental functions;
- estimation of the socioeconomic consequences.

The value of each function is then determined for each landscape unit, calculating a percentage loss of quality and ecosystem functionality over time (Salvati and Zitti, 2005). The total costs of land degradation are obtained by summing these values on all landscape units and environmental functions. This procedure is correct, but it requires well-structured input data, difficult to find even in the most developed countries. Referring to the primary sector, a simplified procedure applicable over large areas, to estimate the potential costs of degradation (Salvati and Carlucci, 2010). It is theoretically connected to the user-cost approach through the evaluation of depletion costs. The purpose is achieved through the integration of different environmental variables with economic information (estimated on a district scale) available for sufficiently long periods of time from statistical sources of national and territorial accounting. The extraction rate of the land capital, in this case, has been estimated by adapting the ESAI index to a pre-defined function.

4.5. Complex Adaptive Systems and System-Level Properties

The procedure to assess socioecological resilience (SER) and community resilience (CR) proposed in LEDDRA comprises three main steps to assess: (i) lower level properties, (ii) system-level properties and (iii) SER and CR. The rationale of this approach draws on the contemporary 'Resilience Thinking' school which focus on the assessment of SER on specific system attributes that play an important role in the dynamics of a SES (Cumming et al., 2005; Anderies et al., 2006; Walker and Salt, 2006). SER and CR are considered as hyper-properties of a SES emerging from system-level properties (namely, Resilience, Adaptability and Transformability, RAT) that derives from lower-level properties. Responses to LEDD modify the characteristics of the SES and, consequently, the LLPs, the system-level properties and SER/CR (Table 2).

Table 2. Schematic presentation of the procedure to assess SER and CR

Socioecological Resilience/Community Resilience	
System-level properties	Resilience - Adaptability - Transformability
Lower level properties (the list is not exhaustive; the properties are arranged in priority order)	Potential available for change - Robustness - Diversity - Redundancy - Connectedness - Modularity - Openness - Flexibility - Continuity - Rapidity - Resourcefulness
Ses characteristics and functions	Characteristics of the components of the three capitals (levels of the three capitals) Relationships among the characteristics of the components of the capitals (SES structure) Multiscale feedback mechanisms

The assessment of lower-level properties (LLPs) is based on the analysis of (i) the characteristics of natural, economic and social capital of a SES, (ii) the relationships among those characteristics that determine the maintenance of its critical ecosystem and socioeconomic functions, and (iii) the critical functions. The assessment of system level properties, more specifically of Resilience, Adaptability and Transformability (RAT), involves the identification of those combinations of LLPs that, through their specific interactions, are considered to shape the 'values' of R, A and T in a given SES. The type assessment differs between past and future periods of a SES. For past periods, RAT is assessed *ex-post*, while for future periods it is assessed *ex ante*. The major difference lies in the fact that while RAT to disturbances that did occur in the past can be objectively assessed, RAT with respect to future disturbances (change) can only be evaluated by means of prospective studies or informed reasoning. There are no specific RAT assessment methods and techniques. The assessment is carried out for each specific study site and relies on the thorough analysis of the SES and deep knowledge of its evolution. A combination of quantitative and qualitative methods and techniques as well as heuristic methods (e.g., expert systems) can be used. The following broad suggestions are based on the pertinent literature.

Resilience can be expressed as a function of those LLPs that relate to the capacity of a SES to stay in a basin of attraction (state), to maintain critical functions and, thus, to secure the provision of ecosystem services. In this perspective, the following four crucial aspects of resilience (Walker et al., 2004) are important to consider in choosing pertinent LLPs: latitude, resistance, precariousness and panarchy (e.g., Briassoulis, 1989). The LLPs that might be first considered in assessing the Resilience of a SES are: potential available for change ('amounts' and types of the components of the three capitals); Robustness; Diversity; Redundancy; Connectedness; and Modularity. For past periods, the analysis focuses on the combinations of the values of these LLPs that explain the observed degree of resilience of the SES. For future periods, it is necessary to consider alternative future situations (scenarios) against which the combinations of the current values of these LLPs that will maintain its resilience should be identified. Specified and general resilience should be assessed.

Adaptability can be expressed as a function of those properties that determine the ability of the SES to manage resilience in the sense of being able to cope with novel situations without losing options for the future; e.g., of maintaining its identity and the critical functions in the process of adaptation to change. According to Walker et al., (2004) ".... adaptability of the system is mainly a function of the social component—the individuals and groups acting to manage the system." Their collective capacity to intentionally manage resilience determines whether they can avoid crossing into an undesirable system regime or succeed in crossing back into a desirable one. Therefore, the LLPs that reflect the adaptability of the SES relate to the four aspects of Resilience mentioned before. In other words, these LLPs relate to the 'value' of the observed or projected resilience. They 'explain' how the social actors influence latitude, resistance, precariousness and panarchy of the state of the SES that result in the observed resilience or will result in the projected resilience.

Transformability can be expressed as a function of those properties that determine the ability of the SES to purpose fully move to another state (basin of attraction or a new stability domain) when ecological, economic, or social (including political) conditions make the existing state untenable (Walker et al., 2004). The emphasis is again put on social and economic components of the SES. Transformation, i.e., creation of new conditions, introduction of new components, changes of the state variables, should ensure the maintenance of critical functions and the provision of ecosystem services. It is likely that the same properties that determine the adaptability of the SES become relevant for assessing its transformability (Walker et al., 2004). The LLPs that might be first considered in assessing the Adaptability of a SES are:

- Potential available for change ('amounts' and types of the components of social and economic capitals)
- Robustness
- Diversity (social and economic)
- Redundancy (social and economic)
- Connectedness (social and economic)
- Modularity
- other (novelty, learning, mode of governance, etc.)

For past periods, the analysis is undertaken only if a major SES transformation has taken place, e.g., from a predominantly agricultural or undeveloped state to a predominantly tourist state. The analysis focuses on the combinations of the 'values' of these LLPs that 'explain' the observed transformation of the SES; i.e., which LLPs were related to low resilience at the current state and which factors triggered the shift to a totally new state under which the SES built up high resilience. For future periods, it is necessary to consider alternative situations (scenarios) - that present an improvement over the current state - to figure out which combinations of values of these LLPs will make the transformation to an alternative future state and secure high resilience of the SES. Walker et al., (2004) note that "it is, nevertheless, likely that there is overlap in the attributes that promote adaptability and transformability. In addition to such common attributes (e.g., diverse and high levels of natural and built capital), we speculate that attributes required for transformability will emphasize novelty, diversity, and organization in human capital—diversity of functional types (kinds of education, expertise, and occupations); trust, strengths, and variety in institutions; speeds and kinds of cross-scale communication, both within the panarchy and between other systems elsewhere."

The assessment of SER (regional level) and CR (community or local level) requires the synthesis of the values of the three system level properties (RAT). It is intended that this assessment essentially concerns the SER and CR of the studied system. For past periods, the synthesis considers the combinations of R, A and T that explain the observed SER of the SES. For future periods, the synthesis suggests the combinations of R, A, and T that will (a) either maintain the SES at the current state if the state is deemed desirable subject to modifications (adaptation) or (b) will steer the SES to a new state if the current state is deemed undesirable.

4.6. Evaluating Responses to Land Degradation

This stage provides the final assessment of the socioecological fit of responses to LEDD. The emphasis of this stage is on the responses to

LEDD that have been identified, assessed and explained. The assessment of fit is basically an evaluation of how the responses to LEDD under consideration have shaped the SER of the SES, based on specific evaluation-of-fit criteria. The procedure followed at this stage comprises the following steps: (i) identification of evaluation-of-fit criteria; (ii) assessment of the fit of (the dominant) responses to LEDD, e.g., evaluation of SER according to specific criteria; (iii) explanation of the fit of responses. The assessment of the socioecological fit of responses to LEDD is carried out for both past, current and future periods.

Sets of 'evaluation-of-fit' criteria can be (a) developed by researchers, reflecting their scientific and personal understanding of the SES and (b) by researchers in cooperation with stakeholders, thus reflecting the experience and value systems of the latter as well as the differences among them. Candidate criteria may concern the environmental (fit in an environmental sense), socio-cultural (fit in a socio-cultural sense) and economic (fit in an economic sense) dimensions of SER. They may include integrated criteria as well. The application of alternative sets of evaluation-of-fit criteria to the multidimensional measures of SER obtained before will obviously produce alternative assessments of the fit of responses to LEDD in a SES. A given response assembly may be fit according to some criteria but not universally, especially if future uncertainty is considered. The final step will be to provide an integrated explanation of the assessment of fit of responses obtained for each group of stakeholders, or evaluation criteria in general, following the complex adaptive systems.

The socioecological fit of responses concerns the degree to which responses contribute to the maintenance (or enhancement) of SER. Maintenance of SER means maintenance of RAT and of the desirable combinations of lower level properties which they emerge from. Maintenance of lower-level properties means securing desirable/proper ranges of values of SES characteristics and maintenance of the critical functions that are necessary for the provision of ecosystem services. Two alternative sets of evaluation-of-fit criteria may be used: (i) Criteria based on lower-level properties; and (ii) Criteria based on critical SES characteristics and functions.

The assessment of the fit of the selected responses to LEDD draws on evaluating how these responses have influenced the SER of the SES and how changes in SER have either maintained or modified the responses to LEDD over time. Evaluation methods and techniques include: (i) simple comparisons of the actual values of the LLPs against (pre-defined) cut-off values of the criteria and use of qualitative scales, combined with informed reasoning, to determine the degree of fit and (ii) multi-criteria evaluation techniques (Pili et al., 2017), intended as heuristic approaches combined with other sources of data and information for a reliable assessment of the fit. It is based on the analysis of the evolution of the SES, the slow and fast variables identified for each state and transition period and the drivers of responses to LEDD in the SES that provide context-specific explanations of how the responses have been producing changes in the LLPs and the SER of the SES and how these changes have been partially or entirely modifying the responses over time.

4.7. Policy Implications

Optimal Response Assemblages (ORA) concerns desirable response assemblages under alternative scenarios. The basic scenario concerns the continuation of the current conditions for a reasonable time horizon (e.g., up to 5 years). The design of ORAs concerns the SES which is analyzed. Therefore, this stage will be carried on for each context to indicate ORAs for that condition. For each future scenario for a SES under study, the assessment of SER and of fit of responses will suggest which SES components and their relationships are responsible for any 'misfit' observed according to criteria and stakeholders considered and, so what characteristics of the SES need to be fixed, if possible, to improve the fit of responses to LEDD for different stakeholder groups. The findings from the analysis of different contexts should be combined with the policy analysis performed at the regional and national level to produce comprehensive and meaningful policy guidelines. Land management and other guidelines will be relevant for the study sites only unless strong justification for their general applicability can be provided.

References

Ali, R. R., and El Baroudy, A. A., (2008). Use of GIS in Mapping the Environmental Sensitivity to Desertification in Wadi El Natrun Depression, Egypt. *Australian Journal of Basic and Applied Sciences*, 2(1), 157-164.

Anderies, J. M., B. H. Walker, and A. P. Kinzig. (2006). Fifteen weddings and a funeral: case studies and resilience-based management. *Ecology and Society* 11(1), 21.

Anselin, L. (1995). Local indicators of spatial association-LISA. *Geographical Analysis,* 27, 93-115.

Antle, J. M., and Heidebrink, G. (1995). Environment and Development: Theory and International Evidence. *Economic Development and Cultural Change*, 43, 603-625.

APAT (2006). *Guidelines for the drafting of project proposals aimed at the implementation of the National Action Plan to combat desertification.* National Agency for Environmental Protection and Technical Services & Desertification Research Unit, University of Sassari. Manuals and Guidelines, Rome.

Arrow, K., Bolin, B., Costanza, R., Dasgupta, P., Folke, C., Holling, C. S., Jansson, B. O., Levin, S., Mäler, K-G., Perrings, C., and Pimentel, D. (1995). Economic Growth, Carrying Capacity, and the Environment. *Science,* 268, 520-521.

Atis, E. (2006). Economic impacts on cotton production due to land degradation in the Gediz Delta, Turkey. *Land Use Policy*, 23, 181-186.

Bajocco, S., De Angelis, A., and Salvati, L. (2012). A satellite-based green index as a proxy for vegetation cover quality in a Mediterranean region. *Ecological Indicators*, 23, 578-587.

Basso, F., Bove, E., Dumontet, S., Ferrara, A., Pisante, M., Quaranta, G., and Taberner, M. (2000). Evaluating environmental sensitivity at the basin scale through the use of geographic information systems and remotely sensed data: an example covering the Agri basin - Southern Italy. *Catena,* 40, 19-35.

Biasi, R., Colantoni, A., Ferrara, C., Ranalli, F., and Salvati, L. (2015). In-between sprawl and fires: Long-term forest expansion and settlement dynamics at the wildland-urban interface in Rome, Italy. *International Journal of Sustainable Development and World Ecology*, 22(6), 467-475.

Biasi, R., Brunori, E., Ferrara, C., and Salvati, L. (2017). Towards sustainable rural landscapes? a multivariate analysis of the structure of traditional tree cropping systems along a human pressure gradient in a mediterranean region. *Agroforestry Systems*, 91(6), 1199-1217.

Bojo, J. (1996). The costs of land degradation in sub-Saharan Africa. *Ecological Economics*, 18, 55-66.

Brandt, J., Geeson, N., and Imeson, A. (2003). *A desertification indicator system for Mediterranean Europe*, DESERTLINKS Project (www.kcl.ac.uk/desertlinks).

Briassoulis, H. (1989). Theoretical orientations in environmental planning: An inquiry into alternative approaches. *Environmental Management*, 13(4), 381-392.

Briassoulis, H. (2011). Governing desertification in Mediterranean Europe: the challenge of environmental policy integration in multi-level governance contexts. *Land Degradation and Development*, 22(3), 313-3.

Brouwer, F. B., Thomas, A. J., and Chadwick, M. J. (1991). *Land-use changes in Europe. Processes of change, environmental transformations and future patterns.* Kluwer Academic Publishers, Dordrecht, The Netherlands.

Camci Cetin, S., Karaca, A., Haktanir, K., and Yand ildiz, H. (2007). Global attention to Turkey due to desertification. *Environmental Monitoring and Assessment,* 128, 489-493.

Carlucci, M., Grigoriadis, E., Rontos, K., and Salvati, L. (2017). Revisiting a Hegemonic Concept: Long-term ‘Mediterranean Urbanization’ in

Between City Re-polarization and Metropolitan Decline. *Applied Spatial Analysis and Policy,* 10(3), 347-362.

Ceccarelli, T., Bajocco, S., Perini, L., and Salvati, L. (2014). Urbanisation and land take of high-quality agricultural soils - Exploring long-term land-use changes and land capability in Northern Italy. *International Journal of Environmental Research*, 8(1), 181-192.

Cimini, D., Tomao, A., Mattioli, W., Barbati, A., and Corona, P. (2013). Assessing impact of forest cover change dynamics on high nature value farmland in Mediterranean mountain landscape. *Annals of Silvicultural Research*, 37(1), 29-37.

Colantoni, A., Ferrara, C., Perini, L., and Salvati, L. (2015a). Assessing trends in climate aridity and vulnerability to soil degradation in Italy. *Ecological Indicators*, 48, 599-604.

Colantoni, A., Mavrakis, A., Sorgi, T., and Salvati, L. (2015b). Towards a 'polycentric' landscape? Reconnecting fragments into an integrated network of coastal forests in Rome. *Rendiconti Lincei*, 26, 615-624.

Colantoni, A., Grigoriadis, E., Sateriano, A., Venanzoni, G., and Salvati, L. (2016). Cities as selective land predators? A lesson on urban growth, deregulated planning and sprawl containment. *Science of the Total Environment*, 545-546, 329-339.

Corona, P., Ascoli, D., Barbati, A., Bovio, G., Colangelo, G., Elia, M., and Lovreglio, R. (2014). Integrated forest management to prevent wildfires under Mediterranean environments. *Annals of Silvicultural Research*, 38(2), 24-45.

Cuadrado-Ciuraneta, S., Durà-Guimerà, A., and Salvati, L. (2017). Not only tourism: unravelling suburbanization, second-home expansion and "rural" sprawl in Catalonia, Spain. *Urban Geography, 38*(1), 66-89.

Cumming, G. S., Barnes, G., Perz, S., Schmink, M., Sieving, K. E., Southworth, J., Binford, M., Holt, R. D., Stickler, C., and Van Holt, T. (2005). An Exploratory Framework for the Empirical Measurement of Resilience. *Ecosystems*, 8, 975-987.

D'Angelo, M., Enne, G., Madrau, S., Percich, L., Previtali, F., Pulina, G., and Zucca, C. (2000). Mitigating land degradation in Mediterranean agro-silvo-pastoral systems: a GIS-based approach. *Catena,* 40, 37-49.

de Groot, R. S., Wilson, M. A., and Bouman, R. M. J. (2002). A typology for the classification, description and valuation of ecosystem functions, goods and services. *Ecological Economics*, 41(3), 393-408.

Delfanti, L., Colantoni, A., Recanatesi, F., Bencardino, M., Sateriano, A., Zambon, I., and Salvati, L. (2016). Solar plants, environmental degradation and local socioeconomic contexts: A case study in a Mediterranean country. *Environmental Impact Assessment Review*, 61, 88-93.

Di Feliciantonio, C., and Salvati, L. (2015). 'Southern' Alternatives of Urban Diffusion: Investigating Settlement Characteristics and Socio-Economic Patterns in Three Mediterranean Regions. *Tijdschrift voor Economische en Sociale Geografie*, 106(4), 453-470.

Duvernoy, I., Zambon, I., Sateriano, A., and Salvati, L. (2018). Pictures from the other side of the fringe: Urban growth and peri-urban agriculture in a post-industrial city (Toulouse, France). *Journal of Rural Studies,* 57, 25-35.

Ferrara, C., Carlucci, M., Grigoriadis, E., Corona, P., and Salvati, L. (2017). A comprehensive insight into the geography of forest cover in Italy: Exploring the importance of socioeconomic local contexts. *Forest Policy and Economics*, 75, 12-22.

Fraser, E. D. G., Dougill, A. J., Mabee, W. E., Reed, M., and McAlpine, P. (2006). Bottom up and top down: Analysis of participatory processes for sustainability indicator identification as a pathway to community empowerment and sustainable environmental management. *Journal of Environmental Management*, 78(2), 114-127.

Garcia Latorre, J., Garcia-Latorre, J., and Sanchez-Picon, A. (2001). Dealing with aridity: socio-economic structures and environmental changes in an arid Mediterranean region. *Land Use Policy*, 18, 53-64.

Hein, L. (2007). Assessing the costs of land degradation: a case study for the Puentes catchment, southeast Spain. *Land Degradation and Development*, 18, 631-642.

Hill, J., Stellmes, M., Udelhoven Th., Roder A., and Sommer, S. (2008). Mediterranean desertification and land degradation: mapping related land-use change syndromes based on satellite observations. *Global and Planetary Change*, 64(3-4), 146-157.

Huby, M., Owen, A., and Cinderby, S. (2007). Reconciling socio-economic and environmental data in a GIS context: an example from rural England. *Applied Geography*, 27, 1-13.

Iosifides, T., and Politidis, T. (2005). Socioeconomic dynamics, local development and desertification in western Lesvos, Greece. *Local Environment,* 10, 487-499.

Kairis, O., Karavitis, C., Kounalaki, A., Salvati, L., and Kosmas, C. (2013). The effect of land management practices on soil erosion and land desertification in an olive grove. *Soil Use and Management*, 29(4), 597-606.

Kairis, O., Kosmas, C., Karavitis, C., Ritsema, C., Salvati, L., Acikalin, S., Alcalá, M., Alfama, P., Atlhopheng, J., Barrera, J., Belgacem, A., Solé-Benet, A., Brito, J., Chaker, M., Chanda, R., Coelho, C., Darkoh, M., Diamantis, I., Ermolaeva, O., Fassouli, V., Fei, W., Feng, J., Fernandez, F., Ferreira, A., Gokceoglu, C., Gonzalez, D., Gungor, H., Hessel, R., Juying, J., Khatteli, H., Khitrov, N., Kounalaki, A., Laouina, A., Lollino, P., Lopes, M., Magole, L., Medina, L., Mendoza, M., Morais, P., Mulale, K., Ocakoglu, F., Ouessar, M., Ovalle, C., Perez, C., Perkins, J., Pliakas, F., Polemio, M., Pozo, A., Prat, C., Qinke, Y., Ramos, A., Ramos, J., Riquelme, J., Romanenkov, V., Rui, L., Santaloia, F., Sebego, R., Sghaier, M., Silva, N., Sizemskaya, M., Soares, J., Sonmez, H., Taamallah, H., Tezcan, L., Torri, D., Ungaro, F., Valente, S., de Vente, J., Zagal, E., Zeiliguer, A., Zhonging, W., and Ziogas, A. (2013b). Evaluation and Selection of Indicators for Land Degradation and Desertification Monitoring: Types of Degradation, Causes, and Implications for Management. *Environmental Management*, 54(5), 971-982.

Kairis, O., Karavitis, C., Salvati, L., Kounalaki, A., and Kosmas, K. (2015). Exploring the impact of overgrazing on soil erosion and land degradation in a dry Mediterranean agro-forest landscape (Crete, Greece). *Arid Land Research and Management,* 29(3), 360-374.

Karamesouti, M., Detsis, V., Kounalaki, A., Vasiliou, P., Salvati, L., and Kosmas, C. (2015). Land-use and land degradation processes affecting soil resources: Evidence from a traditional Mediterranean cropland (Greece). *Catena,* 132, 45-55.

Kazemzadeh-Zow, A., Zanganeh Shahraki, S., Salvati, L., and Samani, N. N. (2017). A spatial zoning approach to calibrate and validate urban growth models. *International Journal of Geographical Information Science*, 31(4), 763-782.

Kosmas, C., Kairis, O., Karavitis, C., Ritsema, C., Salvati, L., Acikalin, S., Alcalá, M., Alfama, P., Atlhopheng, J., Barrera, J., Belgacem, A., Solé-Benet, A., Brito, J., Chaker, M., Chanda, R., Coelho, C., Darkoh, M., Diamantis, I., Ermolaeva, O., Fassouli, V., Fei, W., Feng, J., Fernandez, F., Ferreira, A., Gokceoglu, C., Gonzalez, D., Gungor, H., Hessel, R., Juying, J., Khatteli, H., Khitrov, N., Kounalaki, A., Laouina, A., Lollino, P., Lopes, M., Magole, L., Medina, L., Mendoza, M., Morais, P., Mulale, K., Ocakoglu, F., Ouessar, M., Ovalle, C., Perez, C., Perkins, J., Pliakas, F., Polemio, M., Pozo, A., Prat, C., Qinke, Y., Ramos, A., Ramos, J., Riquelme, J., Romanenkov, V., Rui, L., Santaloia, F., Sebego, R., Sghaier, M., Silva, N., Sizemskaya, M., Soares, J., Sonmez, H., Taamallah, H., Tezcan, L., Torri, D., Ungaro, F., Valente, S., de Vente, J., Zagal, E., Zeiliguer, A., Zhonging, W., and Ziogas, A. (2013). Evaluation and Selection of Indicators for Land Degradation and Desertification Monitoring: Methodological Approach. *Environmental Management*, 54(5), 951-970.

Kosmas, C., Gerontidis, S., and Marathianou, M. (2000). The effect of land-use changes on soil and vegetation over various lithological formations on Lesvos. *Catena,* 40, 51-68.

Kosmas, C., Tsara, M., Moustakas, N., and Karavitis, C. (2003). Identification of indicators for desertification. *Annals of Arid Zones*, 42, 393-416.

Lavado Contador, J. F., Schnabel, S., Gomez Gutiérrez A., and Pulido Fernandez, M. (2009). Mapping sensitivity to land degradation in

Extremadura, SW Spain. *Land Degradation and Development*, 20(2), 129-144.

Loumou, A., Giourga, C., Dimitrakopoulos, P., and Koukoulas, S. (2000). Tourism contribution to agro- ecosystems conservation; the case of Lesbos island, Greece. *Environmental Management*, 26, 363- 370.

Mairota, P., Thornes, J. B., and Geeson, N. (1998). *Atlas of Mediterranean environments in Europe*. The desertification context, Wiley, Chichester.

Makhzoumi, J. M. (1997). The changing role of rural landscapes: olive and carob multi-use tree plantations in the semiarid Mediterranean. *Landscape and Urban Planning,* 37, 115-122.

Marchetti, M., Vizzarri, M., Lasserre, B., Sallustio, L., and Tavone, A. (2015). Natural capital and bioeconomy: challenges and opportunities for forestry. *Annals of Silvicultural Research*, 38(2), 62-73.

Munafò M., Salvati L., and Zitti M. (2013). Estimating soil sealing rate at national level - Italy as a case study. *Ecological Indicators*, 26, 137-140.

Nijkamp, P. (1999). *Environment and regional economics*. In: van der Bergh J. C. J. M. (Ed.) Handbook of environmental and resource economics. Edward Elgar, Cheltenham, UK.

Perez-Trejo, F., and Clark, N. (1996). *Dynamic modelling of complex systems*. In: Brandt C. J., and Thornes J. B. (Eds.). Mediterranean desertification and land-use. Wiley, Chichester.

Pili, S., Grigoriadis, E., Carlucci, M., Clemente, M., and Salvati, L. (2017). Towards sustainable growth? A multi-criteria assessment of (changing) urban forms. *Ecological Indicators*, 76, 71-80.

Puigdefabregas, J., and Mendizabal, T. (1998). Perspectives on desertification: western Mediterranean. *Journal of Arid Environment*, 39, 209-224.

Rontos, K., Grigoriadis, E., Sateriano, A., Syrmali, M., Vavouras, I., and Salvati, L. (2016). Lost in protest, found in segregation: Divided cities in the light of the 2015 "Οχι" referendum in Greece. *City, Culture and Society*, 7(3), 139-148.

Salvati, L. (2013). Monitoring high-quality soil consumption driven by urban pressure in a growing city (Rome, Italy). *Cities,* 31, 349-356.

Salvati, L. (2014). Agro-forest landscape and the 'fringe' city: A multivariate assessment of land-use changes in a sprawling region and implications for planning. *Science of the Total Environment,* 490, 715-723.

Salvati, L., and Carlucci, M. (2010). Estimating land degradation risk for agriculture in Italy using an indirect approach. *Ecological Economics*, 69(3), 511-518.

Salvati, L., and Carlucci, M. (2011). The economic and environmental performances of rural districts in Italy: Are competitiveness and sustainability compatible targets? *Ecological Economics*, 70(12), 2446-2453.

Salvati L., and Ferrara A. (2014). Do land cover changes shape sensitivity to forest fires in peri-urban areas? *Urban Forestry and Urban Greening*, 13(3), 571-575.

Salvati, L., and Zitti, M. (2008a). Assessing the impact of ecological and economic factors on land degradation vulnerability through multiway analysis. *Ecological Indicators*, 9, 357-363.

Salvati, L., Zitti, M., and Ceccarelli, T. (2008b). Integrating economic and environmental indicators in the assessment of desertification risk: A case study. *Applied Ecology and Environmental Research*, 6(1), 129-138.

Salvati, L., and Zitti, M. (2008b). Natural resource depletion and the economic performance of local districts: suggestions from a within-country analysis. *International Journal of Sustainable Development and World Ecology*, 15(6), 518-523.

Salvati, L., and Zitti, M. (2009). Substitutability and weighting of ecological and economic indicators: Exploring the importance of various components of a synthetic index. *Ecological Economics*, 68(4), 1093-1099

Salvati, L., and Zitti, M. (2012). Monitoring vegetation and land-use quality along the rural-urban gradient in a Mediterranean region. *Applied Geography*, 32(2), 896-903.

Salvati, L., Petitta, M., Ceccarelli, T., Perini, L., Di Battista, F., and Scarascia, M. E. V. (2008a). Italy's renewable water resources as estimated because of the monthly water balance. *Irrigation and Drainage*, 57(5), 507-515.

Salvati, L., Perini, L., Sabbi, A., and Bajocco, S. (2012). Climate Aridity and Land-use Changes: A Regional-Scale Analysis. *Geographical Research*, 50, 2, 193-203.

Salvati, L., Tombolini, I., Perini, L., and Ferrara, A. (2013a). Landscape changes and environmental quality: The evolution of land vulnerability and potential resilience to degradation in Italy. *Regional Environmental Change*, 13(6), 1223-1233.

Salvati, L., Morelli, V. G., Rontos, K., and Sabbi, A. (2013b). Latent exurban development: City expansion along the rural-to-urban gradient in growing and declining regions of southern Europe. *Urban Geography*, 34(3), 376-394.

Salvati, L., Sateriano, A., and Zitti, M. (2013c). Long-term land cover changes and climate variations - A country-scale approach for a new policy target. *Land Use Policy,* 30(1), 401-407.

Salvati, L., Sateriano, A., and Grigoriadis, E. (2016). Crisis and the city: profiling urban growth under economic expansion and stagnation. *Letters in Spatial and Resource Sciences,* 9(3), 329-342.

Savo, V., De Zuliani, E., Salvati, L., Perini, L., and Caneva, G. (2012). Long-term changes in precipitation and temperature patterns and their possible impacts on vegetation (Tolfa-Cerite area, central Italy). *Applied Ecology and Environmental Research*, 10(3), 243-266.

Seely, M., and Wohl, H. (2004). Connecting research to combating desertification. *Environmental Monitoring and Assessment*, 99, 23-32.

Sepehr, A., Hassanli, A. M., Ekhtesasi, M. R., and Jamali, J. B. (2007). Quantitative assessment of desertification in south of Iran using MEDALUS method. *Environmental Monitoring and Assessment*, 134, 243-254.

Smiraglia D., Ceccarelli T., Bajocco S., Perini L., and Salvati L. (2015). Unraveling Landscape Complexity: Land Use/Land Cover Changes and Landscape Pattern Dynamics (1954-2008) in Contrasting Peri-

Urban and Agro-Forest Regions of Northern Italy. *Environmental Management*, 56(4), 916-932.

Steer, A. (1998). Making development sustainable. *Advances in Geo-Ecology*, 31, 857-865.

Walker, B. H., Gunderson, L. H., Kinzig, A. P., Folke, C., Carpenter, S. R., and Schultz, L. (2006). A handful of heuristics and some propositions for understanding resilience in social-ecological systems. *Ecology and Society*, 11(1), 13.

Zambon, I., Serra, P., Sauri, D., Carlucci, M., and Salvati, L. (2017). Beyond the 'mediterranean city': Socioeconomic disparities and urban sprawl in three Southern European cities. *Geografiska Annaler, Series B: Human Geography*, 99(3), 319-337.

Zambon, I., Benedetti, A., Ferrara, C., and Salvati, L. (2018). Soil Matters? A Multivariate Analysis of Socioeconomic Constraints to Urban Expansion in Mediterranean Europe. *Ecological Economics*, 146, 173-183.

Zitti, M., Ferrara, C., Perini, L., Carlucci, M., and Salvati, L. (2015). Long-term urban growth and land-use efficiency in Southern Europe: Implications for sustainable land management. *Sustainability* (Switzerland), 7(3), 3359-3385.

In: Land Degradation: The Main Challenge ISBN: 978-1-53615-575-4
Editors: Ilaria Zambon et al.

Chapter 5

CONCLUSION: LAND DEGRADATION AND COMPLEX SOCIOECOLOGICAL SYSTEMS

Ilaria Zambon[1,*], Adele Sateriano[2,†], Luca Salvati[3,‡] and Pavel Cudlín[4,#]

[1]Tuscia University, Viterbo, Italy
[2] Pontificia Universitas Lateranensis, Rome, Italy
[3]Council for Agricultural Research and Economics (CREA), Rome, Italy
[4]Global Change Research Centre, Lipová, Czech Republic

ABSTRACT

Ecosystems can be considered as complex systems where different elements (productive, institutional and contextual) act synergistically on environmental conditions and land degradation processes. Soil degradation, drought, poverty, cultural and technological backwardness

[*] Corresponding Author's E-mail: ilaria.zambon@unitus.it
[†] Author's E-mail: adsateri@tin.it
[‡] Author's E-mail: luca.salvati@crea.gov.it
[#] Author's E-mail: pavel.cudlin@mendelu.cz

are the main causes of degradation of both natural and social environments. Such issues usually affect marginal areas, and this happens both in economically-developed countries and in developing regions. In these areas, sustainable land management is recognized as the element on which to act to improve people's living conditions and safeguard the environment (Salvati and Zitti, 2008b; Salvati and Carlucci, 2011, 2014; Salvati et al., 2013a; Zitti et al., 2015; Biasi et al., 2017; Pili et al., 2017).

Desertification is the most emblematic case of land degradation, the effects of which were first recognized at the beginning of the 20th century (Kosmas et al., 1999, 2003, 2013; Salvati et al., 2009; Kairis et al., 2013a, 2013b). In 1930, most of the Great Plains of the United States of America suffered a prolonged drought which, together with inappropriate agronomic practices, led to soil degradation, which has gone down in history with the term "dust bowls". Specifically, adverse weather and climate conditions appeared that affected the Central United States and Canada between 1931 and 1939 which, leading to soil deterioration, gave rise to sandstorms. This ecological disaster caused an exodus of more than half a million Americans who left their farms in Texas, Kansas and Oklahoma. Only the adoption of more appropriate cultivation methods and the sustainable management of water resources prevented catastrophic consequences in the event of similar droughts. Unfortunately, this has not remained an isolated episode because adverse climatic conditions and poor land management have led to cases of land degradation in almost all areas worldwide (Moonen et al., 2002; Montanarella, 2007; Salvati et al., 2012a; Colantoni et al., 2015a).

Nowadays, global warming, together with the intensification of economic development and population growth, have led to soil degradation, that now affects nearly 40% of the Earth's surface, including some areas of southern Europe (National Committee for the Fight against Desertification, 1998). After experiencing droughts with a general increase in climatic aridity, the Mediterranean basin has in fact been considered one of the most important hotspots for the observation of soil degradation and desertification processes (Kosmas et al., 1999, 2003, 2013; Salvati and Zitti, 2005; Salvati et al., 2009, 2012b; Kairis et al., 2013a, 2013b; Karamesouti et al., 2015; Zambon et al., 2018). It has been widely demonstrated that, in this region, the increasing level of environmental vulnerability is associated with long-term ecological dynamics (e.g., climate aridity, soil deterioration, erosion, salinity and land-use changes) together with socioeconomic, cultural and institutional dynamics that contribute to anthropogenic pressure leading to major landscape transformations (Moonen et al., 2002; Montanarella, 2007; Salvati and Zitti, 2008a; Salvati et al., 2012a; Colantoni et al., 2015a; Di Feliciantonio and Salvati, 2015; Zambon et al., 2017, 2018). All these conditions can be exacerbated by unsustainable land management, especially in fragile areas (Moonen et al., 2002).

5.1. Spatial Complexity of Land Degradation

The integrated analysis of land vulnerability levels at the beginning of the study period and their evolution over time allows identification of a dynamic picture of the environmental conditions that determine, at local and regional level, a higher level of vulnerability and, potentially, a higher risk of desertification (Rubio and Bochet, 1998; Salvati and Zitti, 2009; Salvati and Carlucci, 2010). Depending on anthropogenic pressure and the degree of fragility, Mediterranean land includes a rather heterogeneous set of conditions of vulnerability to soil degradation (Salvati and Carlucci, 2010, 2011; Salvati et al., 2011, 2013b; Munafò et al., 2013; Salvati, 2013; Ceccarelli et al., 2014; Colantoni et al., 2015b). Critical factors include climate conditions (Moonen et al., 2002; Montanarella, 2007; Salvati et al., 2012a; Colantoni et al., 2015a). At end of World War II, quantitative assessments of land degradation depict a geography of land vulnerability based on geographical gradients (Salvati and Zitti, 2012; Salvati et al., 2013a). Empirical analysis allows outlining trend towards the progressive worsening of environmental conditions. Land deterioration presents a rather complex spatial distribution that does not allow the identification of geographical gradients (Salvati and Zitti, 2012; Salvati et al., 2013a). At the local level, however, the level of vulnerability shows more structured trends such as, for example, the growth recorded in coastal areas compared to inland areas. The causes, in this case, can be attributed to climate trends, decrease in rainfall and increase in temperatures (Moonen et al., 2002; Montanarella, 2007; Salvati et al., 2012a; Colantoni et al., 2015a) responsible for the expansion of arid and semi-arid areas (Salvati and Bajocco, 2011) but also to the increase of anthropogenic pressure, landscape change, agricultural intensification and soil salinization (Salvati, 2013). At local level, there are also several areas where the level of vulnerability has declined over the last decades. These are mostly areas with low population density and strong characteristics of rurality and economic marginality, where typical landscapes, traditional agriculture and ancient settlements persist (Salvati and Zitti, 2012; Ceccarelli et al., 2014; Salvati, 2014; Karamesouti et al., 2015; Zitti et al., 2015; Biasi et al.,

2017). Here land abandonment is in some way mitigated by reforestation processes (spontaneous or guided by man) that increase the wooded area (Cimini et al., 2013; Corona et al., 2014; Marchetti et al., 2015; Biasi et al., 2015, 2017; Colantoni et al., 2015b), with positive effects on soil, water and biodiversity (Figure 1).

Figure 1. Anti-desertification sand fences in Morocco. Source: Own elaboration.

Land vulnerability can also be recognized in areas where other types of effects prevail due to predisposing geomorphological characteristics (e.g., soil erosion in badlands, poor vegetation cover). In general, these situations, often associated with inappropriate land management, can easily be observed in Southern Europe but also in other areas of the continent where the same conditions prevail (Bajocco et al., 2012). Conditions of low vulnerability, which have persisted so throughout the period examined, have been found in limited areas where ecosystems have not been compromised by human pressure and climate aggressiveness (Moonen et al., 2002; Montanarella, 2007; Salvati et al., 2012a; Colantoni et al., 2015a).

Scenarios analysis may confirm the trends observed over the last decades and highlights worse environmental factors predisposing land degradation. This trend can be attributed in part to the extension of semi-arid areas due to the local decrease in precipitation, and in part to an increase in anthropogenic pressure which is mainly reflected in land-use changes (Kosmas et al., 2000b; Tanrivermis, 2003; Salvati and Zitti, 2012; Ceccarelli et al., 2014; Salvati, 2014; Zitti et al., 2015). The negative consequences can also be seen in terms of landscape transformations which, thanks to the growing infrastructure and artificialization of the territory due to local processes of agricultural, industrial and tourism intensification, has affected man-made areas, such as coastal, flat and peri-urban districts. To give a relevant example, one of the most significant changes observed in Europe (Turri, 1999) is the transformation of the landscape of flat areas due to urban sprawl. This phenomenon has led to the conversion of agricultural land to peri-urban settlements through a process of progressive sealing of the soil that has grown significantly from the 1960s to the present days (Munafò et al., 2013).

This evindence indicates the importance of effective mitigation policies with a paradigm shift in operational approach, considering together rural, economically and socially disadvantaged areas and peri-urban districts in rapid transition where the level of land vulnerability appears to be rising sharply (Duvernoy et al., 2018). According to Briassoulis (2011), effective measures against land degradation and consistent with the concept of sustainable development should go beyond the narrow sectoral vision to pursue a truly multi-target and multi-scalar approach (Salvati and Zitti, 2008b; Salvati and Carlucci, 2011, Zitti et al., 2015, 2014; Biasi et al., 2017; Pili et al., 2017). This is particularly important when different degradation processes act synergistically at the local level with interactions and feedback effects that are difficult to manage by environmental measures in the sector. The 'local' dimension therefore remains a fundamental prerequisite for any intervention strategy (and should include specific actions to mitigate territorial disparities).

5.2. The Regional Context

There has been a profound change in the relationship between man and the environment in recent decades, mainly due to rapid and widespread economic development and the emergence of new patterns of consumption (Salvati, 2013). The agricultural sector - as a possible interface between productive interests and environmental protection - is an emblematic case, even if with characteristics that distinguish it from other productive activities. Environmental problems have been contrasted with a growing awareness of the impact of specific choices of social actors. The gradual disappearance of traditional landscapes, the intensive use of natural resources in fertile areas and the tendency to abandon agricultural land where economic convenience is lacking, reflects the inherent transformations that have taken place in the relationship between agricultural activity and the environment, representing latent forms of land degradation (Duvernoy et al., 2018). However, in recent years, even in the "rural world", a greater awareness in the use of agricultural and livestock practices with high environmental pressure has been consolidated in favor of a land management more inclined to preserve and safeguard natural resources and landscapes (Salvati, 2013). In this perspective, production processes have assumed a local importance with significant consequences, such as the growing role of the active involvement of economic, social and institutional actors of the territory. In fact, the local Administrations are today in charge of policies to advance the production systems which, thanks to decentralization processes, sees Regional authorities more involved in land management and reduction of social inequalities, through the involvement of local finance, education centers and the entrepreneurial system.

The introduction of the concept of sustainable development, understood as balanced in all components (economic, social, environmental) in time and space (Zuindeau, 2007), is a useful contribution to this process, not only because this notion of sustainability is independent of the specific definition of development but also because of the possibility of identifying policy and feedback objectives and mechanisms that incorporate the principle of sustainability (Salvati and Zitti, 2008b; Salvati and Carlucci, 2011, 2014; Zitti et al., 2015; Biasi et al., 2017; Pili et al., 2017). The possibility that these expectations materialize in economic policy measures is complex to analyze, but it is linked to the relationship between the time horizon of policy makers and the effects of unsustainability.

In such a socioeconomic context, however, the development path can generate a negative trade-off with the environment due to the more impactful production components. The development path, however, should stimulate responses to contrast and mitigate environmental degradation in these cases. Land degradation does not have a common profile at the national level, being characterized by different factors of production, institutional and value contexts at the basis of territorial disparities. It should also be noted that not only the environmental mechanisms of degradation, but also the channels of transmission of economic and social impacts on the ecosystem differ within and between region.

5.3. Rethinking Complexity: Strategies Against Desertification

The problems resulting from land degradation raise questions about its causal relationships, going beyond conventional assumptions. According to scientific literature, relevant hypotheses representing the complex interrelationships underlying land degradation were proposed in earlier studies. Obviously, under specific socioeconomic conditions, other hypotheses are plausible, but they appear to be of less interest due to the implications of policy on a supra national scale (Briassoulis, 2005; Wilson and Juntti, 2005):

- The 'Malthusian' hypothesis. Although empirical evidence provides limited support for the hypothesis that population growth and spatial concentration are causal factors for degradation, population density remains a key (albeit indirect) determinant of environmental pressure. The weight of each component of this causal chain still needs to be examined in depth to identify correct mitigation policies.

- The hypothesis of the 'Kuznets Environmental Curve'. The inverted U-shaped relationship between land degradation indicators and per capita income level has not yet been sufficiently verified, even if some evidence of empirical relations (linear and significant) between land vulnerability and per capita income are documented (Salvati and Bajocco 2011). According to the interpretation of land 'policies' in response to degradation, there is a clear and automatic relationship between income growth and environmental protection through proper land management. Also, this automatism is not fully shared, it is evident the implication of other variables whose involvement needs specific insights (for example, the role of technological evolution, the impact of credit markets, the mitigation of social conflicts, short and medium range migrations).
- The hypothesis of the 'poverty trap'. To date, there is little experimental evidence to demonstrate the link between land degradation and poverty. In the countries of the Mediterranean area, particularly the role of this factor should be evaluated in connection with other processes, especially because of the economic growth that took place after World War II. In the light of the most recent economic crisis that is causing, from 2010 onwards (Salvati et al., 2016), a widespread and serious impoverishment of populations on a global scale, it would be interesting to explore this relationship however this term is particularly difficult because of poor availability of updated statistical data.
- The hypothesis of the influence of the 'land structure'. Fragmentation of ownership and spread of tenancy, although considered relevant factors in developing countries, can have contradictory or poorly documented effects on degradation processes in Mediterranean Europe (Salvati and Zitti, 2005; Salvati et al., 2012b; Karamesouti et al., 2015; Zambon et al., 2018). This

is clear from the traditional view that property titles contribute to the preservation of natural resources as local actors (farmers) are oriented to pursue sustainable production policies. This hypothesis should be supported with empirical evidence.

- The hypothesis of "agricultural intensification". Higher returns from intensive land-use would encourage the cultivation of new land, attract new labor, but at the same time profitability could be negatively affected by the increased supply of products. Intensification, when it does not result in a reduction of impacts, becomes synonymous with land degradation because of the pressure exerted on the environment, thanks to the most sophisticated technologies, leading to over-exploitation or depletion of natural resources mainly water and soil.
- The hypothesis of 'territorial disparities'. Inappropriate land management, particularly regarding access to natural resources, can lead to socioeconomic disparities and become a catalyst for land degradation. However, the extent to which the starting socioeconomic context can influence processes of economic polarization and the extent to which social inequalities can have an impact on desertification processes is not yet fully understood (Rontos et al., 2016; Carlucci et al., 2017; Zambon et al., 2017, 2018).

In conclusion, according to Blaikie and Brookfield (1987), land degradation (and desertification in the broadest sense) can be considered a typical socioeconomic problem: productive land-use and land management are essentially derived from social needs (Zambon et al., 2018). Therefore, soil productivity, the carrying capacity of ecosystems, conflicts for natural resources and, ultimately, the pursuit of sustainable development, are the products of the dynamic interaction between man and nature (Mainguet, 1994; Steer 1998; Conacher, 2000; Arshad and Martin, 2002; Salvati and Zitti, 2008b; Salvati and Carlucci, 2011; Karamesouti et al., 2015; Zitti et al., 2015; Biasi et al., 2017; Pili et al., 2017). These issues should be appropriately referred to the notion of sustainable development and, in relation to this, must be interpreted and implemented (Singh and Singh, 1995; Chopra and Gulati, 1997; Barbier, 2000; Johnson and Lewis, 2007; Salvati and Zitti, 2008b; Salvati and Carlucci, 2014; Zitti et al., 2015; Biasi et al., 2017; Pili et al., 2017). The hypothesis underlying desertification also confirms the need for a holistic approach to sustainable development (Borzel, 2000), otherwise, the causal chains described above could turn into environmental spirals that are difficult to counteract.

Intimate linkages between land degradation and human pressure stimulate environmental measures able to address the new issues of spatial planning, or a progressive rebalance of population (Salvati and Zitti, 2012; Salvati et al., 2013a; Salvati, 2013; Carlucci et al., 2017). In this regard, internal migrations have contributed to more recent pattern of urban sprawl (Salvati et al., 2013a, 2013c; Salvati, 2014; Morelli et al., 2014; Colantoni et al., 2016; Cuadrado-Ciuraneta et al., 2017; Zambon et al., 2017).

From this point of view, research is of great importance because of the concrete possibility of indicating policy responses to be implemented specifically in the various environments (Verstraete et al., 2008), considering the institutional framework, the range of possible strategies to be pursued and their interactions with the biophysical and socioeconomic dimensions of desertification (Boardman et al., 2003; Salvati and Zitti, 2005; Anthopoulou et al., 2005; Di Feliciantonio and Salvati, 2015; Zambon et al., 2017, 2018). In this sense, it has been demonstrated how desertification processes can have measurable effects on the reduction of land resources in the Mediterranean area (Tanrivermis, 2003; Atis, 2006; Hein, 2007).

Development policies should be more effectively managed with the effect of endogenous and exogenous factors of degradation (Bojo, 1996; Fernandez, 2002; Romm, 2011). For this reason, it is necessary that the current concept of desertification will be supplemented with a new paradigm where the socioeconomic dimension assumes a pivotal role (Wilson and Juntti, 2005; Di Feliciantonio and Salvati, 2015; Zambon et al., 2017, 2018). While, land degradation represents a serious social problem with obvious repercussions on productive structures and ecosystem functionality, policy approaches should consider this new systemic approach. This could facilitate the achievement of the desired objectives by competent institutions in the fight against land deterioration at the appropriate scale of intervention.

REFERENCES

Anthopoulou, B., Panagopoulos, A., and Karyotis, T. (2005). The impact of land degradation on landscape in northern Greece. *Landslides,* 3, 289-294.

Arshad, M.A., and Martin, S. (2002). Identifying critical limits for soil quality indicators in agro-ecosystems. *Agriculture, Ecosystems Environment*, 88(2), 153-160.

Atis, E. (2006). Economic impacts on cotton production due to land degradation in the Gediz Delta, Turkey. *Land Use Policy*, 23, 181-186.

Bajocco, S., De Angelis, A., and Salvati, L. (2012). A satellite-based green index as a proxy for vegetation cover quality in a Mediterranean region. *Ecological Indicators*, 23, 578-587.

Barbier, E.B. (2000). The economic linkages between rural poverty and land degradation: some evidence from Africa. *Agriculture, Ecosystems and Environment*, 82(1-3), 355-370.

Basso, F., Bove, E., Dumontet, S., Ferrara, A., Pisante, M., Quaranta, G., and Taberner, M. (2000). Evaluating environmental sensitivity at the basin scale through the use of geographic information systems and remotely sensed data: an example covering the Agri basin - Southern Italy. *Catena,* 40, 19-35.

Biasi, R., Brunori, E., Ferrara, C., and Salvati, L. (2017). Towards sustainable rural landscapes? a multivariate analysis of the structure of traditional tree cropping systems along a human pressure gradient in a mediterranean region. *Agroforestry Systems*, 91(6), 1199-1217.

Biasi, R., Colantoni, A., Ferrara, C., Ranalli, F., and Salvati, L. (2015). In-between sprawl and fires: Long-term forest expansion and settlement dynamics at the wildland-urban interface in Rome, Italy. *International*

Journal of Sustainable Development and World Ecology, 22(6), 467-475.

Blaikie, P., and Brookfield, H.C. (1987). *Land degradation and society*. Methuen, London.

Boardman, J., Poesen, J., and Evans, R. (2003). Socioeconomic factors in soil erosion and conservation. *Environment Science & Policy*, 6, 1-6.

Bojo, J. (1996). The costs of land degradation in sub-Saharan Africa. *Ecological Economics*, 18, 55-66.

Borzel, T.A. (2000). Why there is no southern problem: on environmental leaders and laggards in the European Union. *Journal of European Public Policies,* 7(1), 141-162.

Briassoulis, H. (2005). *Policy integration for complex environmental problems.* Ashgate, Aldershot.

Briassoulis, H. (2011). Governing desertification in Mediterranean Europe: the challenge of environmental policy integration in multi-level governance contexts. *Land Degradation and Development*, 22(3), 313-3.

Brouwer, F.B., Thomas, A.J., and Chadwick, M.J. (1991). *Land-use changes in Europe. Processes of change, environmental transformations and future patterns.* Kluwer Academic Publishers, Dordrecht, The Netherlands.

Carlucci, M., Grigoriadis, E., Rontos, K., and Salvati, L. (2017). Revisiting a Hegemonic Concept: Long-term 'Mediterranean Urbanization' in Between City Re-polarization and Metropolitan Decline. *Applied Spatial Analysis and Policy*, 10(3), 347-362.

Ceccarelli, T., Bajocco, S., Perini, L.L., and Salvati, L. (2014). Urbanisation and land take of high-quality agricultural soils - Exploring long-term land-use changes and land capability in Northern Italy. *International Journal of Environmental Research*, 8(1), 181-192.

Chopra, K., and Gulati, S.C. (1997). Environmental degradation and population movements: the role of property rights. *Environmental Resource Economics,* 9, 383-408.

Cimini, D., Tomao, A., Mattioli, W., Barbati, A., and Corona, P. (2013). Assessing impact of forest cover change dynamics on high nature value farmland in Mediterranean mountain landscape. *Annals of Silvicultural Research*, 37(1), 29-37.

Colantoni, A., Ferrara, C., Perini, L., and Salvati, L. (2015b). Assessing trends in climate aridity and vulnerability to soil degradation in Italy. *Ecological Indicators,* 48, 599-604.

Colantoni, A., Grigoriadis, E., Sateriano, A., Venanzoni, G., and Salvati, L. (2016). Cities as selective land predators? A lesson on urban growth, deregulated planning and sprawl containment. *Science of the Total Environment*, 545-546, 329-339.

Colantoni, A., Mavrakis, A., Sorgi, T., and Salvati, L. (2015a). Towards a 'polycentric' landscape? Reconnecting fragments into an integrated network of coastal forests in Rome. *Rendiconti Lincei*, 26, 615-624.

Conacher, A.J. (2000). *Land degradation.* Kluwer Academic Publishers, Dordrecht.

Conacher, A.J., and Sala, M. (eds.) (1998). *Land degradation in Mediterranean environments of the world.* Wiley, Chichester.

Cuadrado Ciurancta, S., Durà-Guimerà, A., and Salvati, L. (2017). Not only tourism: unravelling suburbanization, second-home expansion and 'rural' sprawl in Catalonia, Spain. *Urban Geography*, 38(1), 66-89.

Di Feliciantonio, C., and Salvati, L. (2015). 'Southern' Alternatives of Urban Diffusion: Investigating Settlement Characteristics and Socio-Economic Patterns in Three Mediterranean Regions. *Tijdschrift voor Economische en Sociale Geografie*, 106(4), 453-470.

Duvernoy, I., Zambon, I., Sateriano, A., and Salvati, L. (2018). Pictures from the other side of the fringe: Urban growth and peri-urban agriculture in a post-industrial city (Toulouse, France). *Journal of Rural Studies*, 57, 25-35.

Hein, L. (2007). Assessing the costs of land degradation: a case study for the Puentes catchment, southeast Spain. *Land Degradation and Development*, 18, 631-642.

Johnson, D.L., and Lewis, L.A. (2007). *Land degradation - Creation and destruction*, Rowman & Littlefield, Inc. Lahnam, Maryland.

Kairis, O., Karavitis, C., Kounalaki, A., Salvati, L., and Kosmas, C. (2013). The effect of land management practices on soil erosion and land desertification in an olive grove. *Soil Use and Management*, 29(4), 597-606.

Karamesouti, M., Detsis, V., Kounalaki, A., Vasiliou, P., Salvati, L., and Kosmas, C. (2015). Land-use and land degradation processes affecting soil resources: Evidence from a traditional Mediterranean cropland (Greece). *Catena,* 132, 45-55.

Kosmas, C., Kairis, O., Karavitis, C., Ritsema, C., Salvati, L., Acikalin, S., Alcalá, M., Alfama, P., Atlhopheng, J., Barrera, J., Belgacem, A., Solé-Benet, A., Brito, J., Chaker, M., Chanda, R., Coelho, C., Darkoh, M., Diamantis, I., Ermolaeva, O., Fassouli, V., Fei, W., Feng, J., Fernandez, F., Ferreira, A., Gokceoglu, C., Gonzalez, D., Gungor, H., Hessel, R., Juying, J., Khatteli, H., Khitrov, N., Kounalaki, A., Laouina, A., Lollino, P., Lopes, M., Magole, L., Medina, L., Mendoza, M., Morais, P., Mulale, K., Ocakoglu, F., Ouessar, M., Ovalle, C., Perez, C., Perkins, J., Pliakas, F., Polemio, M., Pozo, A., Prat, C., Qinke, Y., Ramos, A., Ramos, J., Riquelme, J., Romanenkov, V., Rui, L., Santaloia, F., Sebego, R., Sghaier, M., Silva, N., Sizemskaya, M., Soares, J., Sonmez, H., Taamallah, H., Tezcan, L., Torri, D., Ungaro, F., Valente, S., de Vente, J., Zagal, E., Zeiliguer, A., Zhonging, W., and Ziogas, A. (2013). Evaluation and Selection of Indicators for Land Degradation and Desertification Monitoring: Methodological Approach. *Environmental Management*, 54(5), 951-970.

Kosmas, C., Gerontidis, S., and Marathianou, M. (2000). The effect of land-use changes on soil and vegetation over various lithological formations on Lesvos, *Catena*, 40, 51-68.

Kosmas, C., Kirkby, M., and Geeson, N. (1999). *Manual on key indicators of desertification and mapping environmental sensitive areas to desertification.* European Commission, Directorate General, Project ENV4 CT 95 0119, EUR 18882, Bruxelles (http://www.kcl.ac.uk/projects/desertlinks/downloads/publicdownloads/ESA%20Manual.pdf).

Kosmas, C., Tsara, M., Moustakas, N., and Karavitis, C. (2003). Identification of indicators for desertification. *Annals of Arid Zones*, 42, 393-416.

Mainguet, M. (1994). *Desertification: natural background and human mismanagement.* Springer, Berlin.

Marchetti, M., Vizzarri, M., Lasserre, B., Sallustio, L., and Tavone, A. (2015). Natural capital and bioeconomy: challenges and opportunities for forestry. *Annals of Silvicultural Research*, 38(2), 62-73.

Montanarella, L. (2007). Trends in land degradation in Europe. In: Sivakumar M.V., N'diangui, N. (Eds), *Climate and land degradation.* Springer, Berlin.

Moonen, A.C., Ercoli, L., Mariotti, M., and Masoni, A. (2002). Climate change in Italy indicated by agrometeorological indices over 122 years. *Agricultural and Forest Meteorology,* 111, 13-27.

Morelli, V.G., Rontos, K., and Salvati, L. (2014). Between suburbanisation and re-urbanisation: revisiting the urban life cycle in a Mediterranean compact city. *Urban Research and Practice*, 7(1), 74-88

Munafò, M., Salvati, L., and Zitti, M. (2013). Estimating soil sealing rate at national level - Italy as a case study. *Ecological Indicators*, 26, 137-140.

Pili, S., Grigoriadis, E., Carlucci, M., Clemente, M., and Salvati, L. (2017). Towards sustainable growth? A multi-criteria assessment of (changing) urban forms. *Ecological Indicators*, 76, 71-80.

Romm, J. (2011). Desertification: The next dust bowl. *Nature,* 478, 450-451.

Rontos, K., Grigoriadis, E., Sateriano, A., Syrmali, M., Vavouras, I., and Salvati, L. (2016). Lost in protest, found in segregation: Divided cities in the light of the 2015 'Οχι' referendum in Greece. *City, Culture and Society*, 7(3), 139-148.

Rubio, J.L., Safriel, U., Blum, W.E.H., and Pedrazzini, F. (2009). *Water scarcity, land degradation and desertification in the Mediterranean region*. Springer, Heidelberg.

Salvati, L. (2013). Monitoring high-quality soil consumption driven by urban pressure in a growing city (Rome, Italy). *Cities*, 31, 349-356.

Salvati, L. (2014). Agro-forest landscape and the 'fringe' city: A multivariate assessment of land-use changes in a sprawling region and implications for planning. *Science of the Total Environment*, 490, 715-723.

Salvati, L., and Bajocco, S. (2011). Land sensitivity to desertification across Italy: past, present, and future. *Applied Geography*, 31(1), 223-231.

Salvati, L., and Carlucci, M. (2010). Estimating land degradation risk for agriculture in Italy through an indirect approach. *Ecological Economics*, 69, 511-518.

Salvati, L., and Carlucci, M. (2011). The economic and environmental performances of rural districts in Italy: Are competitiveness and sustainability compatible targets? *Ecological Economics*, 70(12), 2446-2453.

Salvati, L., and Carlucci, M. (2014). A composite index of sustainable development at the local scale: Italy as a case study. *Ecological Indicators*, 43, 162-171.

Salvati, L., and Zitti, M. (2005). Land degradation in the Mediterranean Basin: Linking bio-physical and economic factors into an ecological perspective. *Biota,* 6, 43132, 67-77.

Salvati, L., and Zitti, M. (2008a). Assessing the impact of ecological and economic factors on land degradation vulnerability through multiway analysis. *Ecological Indicators*, 9, 357-363.

Salvati, L., and Zitti, M. (2008b). Natural resource depletion and the economic performance of local districts: suggestions from a within-country analysis. *International Journal of Sustainable Development and World Ecology*, 15(6), 518-523.

Salvati, L., and Zitti, M. (2009). Substitutability and weighting of ecological and economic indicators: Exploring the importance of various components of a synthetic index. *Ecological Economics*, 68(4), 1093-1099

Salvati, L., and Zitti, M. (2012). Monitoring vegetation and land-use quality along the rural-urban gradient in a Mediterranean region. *Applied Geography*, 32(2), 896-903.

Salvati, L., Zitti, M., Ceccarelli, T., and Perini, L. (2009). Developing a synthetic index of land vulnerability to drought and desertification. *Geographical Research,* 47(3), 280-291.

Salvati, L., Bajocco, S., Ceccarelli, T., Zitti, M., and Perini, L. (2011). Towards a process-based evaluation of land vulnerability to soil degradation in Italy. *Ecological Indicators*, 11(5), 1216-1227.

Salvati, L., Gemmiti, R., and Perini, L. (2012b). Land degradation in Mediterranean urban areas: An unexplored link with planning? *Area,* 44(3), 317-325.

Salvati, L., Perini, L., Sabbi, A., and Bajocco, S. (2012a). Climate Aridity and Land-use Changes: A Regional-Scale Analysis. *Geographical Research*, 50(2), 193-203.

Salvati, L., Morelli, V.G., Rontos, K., and Sabbi, A. (2013a). Latent exurban development: City expansion along the rural-to-urban gradient in growing and declining regions of southern Europe. *Urban Geography*, 34(3), 376-394.

Salvati, L., Tombolini, I., Perini, L., and Ferrara, A. (2013b). Landscape changes and environmental quality: The evolution of land vulnerability and potential resilience to degradation in Italy. *Regional Environmental Change*, 13(6), 1223-1233.

Salvati, L., Zitti, M., and Sateriano, A. (2013c). Changes in city vertical profile as an indicator of sprawl: Evidence from a Mediterranean urban region. *Habitat International*, 38, 119-125.

Salvati, L., Sateriano, A., and Grigoriadis, E. (2016). Crisis and the city: profiling urban growth under economic expansion and stagnation. *Letters in Spatial and Resource Sciences*, 9(3), 329-342.

Singh, J., and Singh, J.P. (1995). Land degradation and economic sustainability, *Ecological Economics*, 15, 77-86.

Steer, A. (1998). Making development sustainable. *Advances in Geo-Ecology*, 31, 857-865.

Tanrivermis, H. (2003). Agricultural land-use change and sustainable use of land resources in the Mediterranean region of Turkey. *Journal of Arid Environment*, 54, 553-564.

Turri, E. (1999). *La megalopoli padana* [*The galaxies are dancing*]. Marsilio, Venezia.

Verstraete, M.M., Brink, A.B., Scholes, R.J., Beniston, M., and Stafford Smith, M. (2008). Climate change and desertification: Where do we stand, where should we go? *Global and Planetary Change*, 64(3-4), 105-110.

Wilson, G.A., and Juntti, M. (2005). *Unravelling desertification: policies and actor networks in Southern Europe*, Wageningen, Wageningen Academic Publishers.

Zambon, I., Serra, P., Sauri, D., Carlucci, M., and Salvati, L. (2017). Beyond the 'mediterranean city': Socioeconomic disparities and urban sprawl in three Southern European cities. *Geografiska Annaler, Series B: Human Geography*, 99(3), 319-337.

Zambon, I., Benedetti, A., Ferrara, C., and Salvati, L. (2018). Soil Matters? A Multivariate Analysis of Socioeconomic Constraints to Urban Expansion in Mediterranean Europe. *Ecological Economics*, 146, 173-183.

Zitti, M., Ferrara, C., Perini, L., Carlucci, M., and Salvati, L. (2015). Long-term urban growth and land-use efficiency in Southern Europe: Implications for sustainable land management. *Sustainability* (Switzerland), 7(3), 3359-3385.

Zuindeau, B. (2007). Territorial equity and sustainable development. *Environmental Values*, 16, 253-268.

Contributors

Editors/Chapter Authors

Ilaria Zambon
Department of Agricultural and Forestry scieNcEs (DAFNE),
Tuscia University, Viterbo, Italy

Luca Salvati
Council of Agricultural Research and Economics (CREA),
Research Centre for Forestry and Wood, Arezzo, Italy

Carlotta Ferrara
Council of Agricultural Research and Economics (CREA),
Research Centre for Forestry and Wood, Arezzo, Italy

Chapter Authors

Rita Biasi
Department for Innovation in Biological, Agro-food
and Forest Systems (DIBAF), Tuscia University,
Viterbo, Italy

Andrea Colantoni
Department of Agricultural and Forestry scieNcEs (D.A.F.N.E.),
Tuscia University, Viterbo, Italy

Massimo Cecchini
Department of Agricultural and Forestry scieNcEs,
Tuscia University, Viterbo, Italy

Pavel Cudlín
Global Change Research Centre,
Academy of Sciences of the Czech Republic, Lipová, Czech Republic

Anastasios Mavrakis
Institute of Urban Environment and Human Resources,
Department of Economic and Regional Development,
Panteion University, Athens, Greece

Kostas Rontos
Sociology Department, University of the Aegean,
Mytilini, Greece

Adele Sateriano
Council of Agricultural Research and Economics (CREA),
Research Centre for Forestry and Wood, Arezzo, Italy

Pere Serra
Department of Geography, Autonomous University of Barcelona,
Barcelona, Spain

INDEX

D

E

F

G

H

I

K

L

M

N

O

P

R

S

T

U

V

W

Related Nova Publications

A Closer Look at Climate Change

Editor: Reggie Paredes

Series: Climate Change and its Causes, Effects and Prediction

Book Description: In *A Closer Look at Climate Change*, referring to a philosophy of science called critical realism, the authors explain why evaluating climate change adaptation interventions is challenging. Critical realism has been utilized in various social science disciplines such as economics, sociology, geography and ecology.

Softcover ISBN: 978-1-53614-600-4
Retail Price: $82

Climate Change Effects on Soils: Aspects and Considerations

Editor: Claudia Holmes

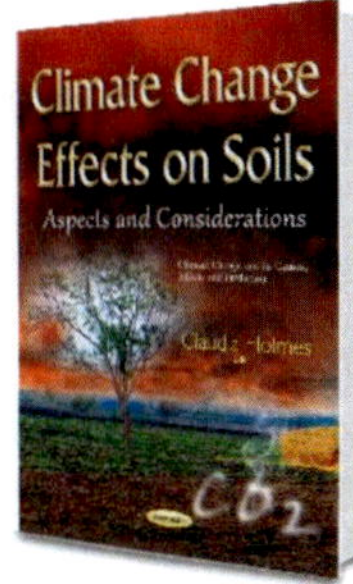

Series: Climate Change and its Causes, Effects and Prediction

Book Description: The objective of this book is to initiate and further stimulate a discussion about some important and challenging aspects of climate-change effects on soils, such as accelerated weathering of soil minerals and resulting C and elemental fluxes in and out of soils, soil/geo-engineering methods used to increase C sequestration in soils, soil organic matter (SOM) protection, transformation and mineralization, and SOM temperature sensitivity.

Softcover ISBN: 978-1-63482-773-7
Retail Price: $69

To see complete list of Nova publications, please visit our website at www.novapublishers.com